AF680120

10 Oklahoma OSTP Grade 8 Math Practice Tests

The Ultimate Test Prep Collection with Answer Explanations

Dr. A. Nazari

Published by View Math Education

ViewMath.com

10 Practice Tests

Grade 8 Mathematics

Welcome!

*This book contains **10 full-length practice tests** — the most comprehensive preparation you can get for your Grade 8 math assessment. Each test covers all six topics:*

- *Irrational Numbers*
- *Powers & Scientific Notation*
- *Linear Equations*
- *Functions*
- *Geometry*
- *Data & Relationships*

Ten tests give you the practice needed to walk into the real test feeling fully prepared.

Thorough preparation leads to outstanding results.

> *“Ten full tests! By the time you finish, there won't be any surprises on test day.”*

How to Use This Book

A complete 10-test preparation program

What's Inside

- ***10 Full-Length Practice Tests*** — *each covers all 6 chapters of Grade 8 math: irrational numbers, exponents & scientific notation, linear equations, functions, geometry, and data analysis.*
- ***Detailed Answer Explanations*** — *every question includes a step-by-step solution so you learn from every mistake.*
- ***Formula Reference Sheet*** — *all the key Grade 8 formulas you need, organized and ready for quick review.*
- ***Test Tracker*** — *log your scores across all 10 tests and monitor your progress from start to finish.*

Your 10-Test Training Plan

★ PHASE 1: Foundation (Tests 1–3)

Untimed or soft-timed. Focus on understanding the format, identifying strengths and weaknesses, and building good study habits.

★★ PHASE 2: Building Skills (Tests 4–7)

Timed (70 minutes each). Work on pacing, accuracy, and showing complete solutions. Review weak topics between tests.

★★★ PHASE 3: Test-Day Ready (Tests 8–10)

Full test conditions: strict timing, quiet space, no notes. Compare scores with your early tests to see your growth.

Schedule: *Take one test every 3–4 days, or one per week. Use study days between tests to review.*

Types of Questions

Multiple Choice: *Four options — work the problem first, then match. Eliminate obviously wrong answers to narrow your choices.*

Short Answer & Constructed Response: *Show every step: equations, substitutions, simplifications. Partial credit rewards correct reasoning even if the final answer is off.*

Graphing & Data Analysis: *Plot points, draw lines, interpret graphs. Label axes clearly.*

Tip: *Ten tests is a full preparation program. Don't rush. The key is what you do between tests — study, review, and understand your mistakes before moving forward.*

Test-Taking Tips

Your complete test-day toolkit

Before the Test

- *Review your notes from the previous test — focus on your weak topics*
- *Set up a quiet, clean workspace with all your materials ready*
- *Start with a positive mindset: you've prepared for this*

During the Test

- *Read each problem fully before calculating anything*
- *Write the formula or set up the equation first, then substitute values*
- *Show all your work — every step, every operation*
- *If stuck for more than 2 minutes, mark it and move on*
- *Use estimation to check if your answers are reasonable*

After the Test

- *Read the full explanation for every question you got wrong*
- *Write down which topics gave you trouble (not just question numbers)*
- *Study those topics before taking the next test*
- *Record your score in the Test Tracker*

Common Mistakes in Grade 8 Math

Exponents: $(ab)^n = a^n b^n$, but $a^m + a^n \neq a^{m+n}$. Only multiply/divide to combine.

Slope formula: $m = \frac{y_2 - y_1}{x_2 - x_1}$ — keep the order consistent.

Systems of equations: The solution must satisfy both equations.

Transformations: Rotations and reflections change position; dilations change size.

Volume: Use $\pi \approx 3.14$ or leave as π — match what the question asks.

“The students who improve the most aren't the ones who take the most tests — they're the ones who carefully review every mistake. Make that your priority.”

What You'll Need

Gather these materials before you begin

Materials Checklist

Sharpened Pencils *— #2 pencils, at least two*

Good Eraser *— for clean corrections*

Scratch Paper *— for working out problems*

Ruler / Straightedge *— for graphing & geometry*

Quiet Space *— no distractions*

Focused Mind *— ready to do your best*

Allowed Materials

- ✔ *Pencils and eraser*
- ✔ *Scratch paper (provided on official test day)*
- ✔ *Ruler or straightedge (if required)*
- ✔ *Protractor (if required)*

Not Allowed

- ✖ *Calculator (unless your state test allows it)*
- ✖ *Cell phone or any electronic device*
- ✖ *Notes, textbooks, or reference sheets*
- ✖ *Help from others during the test*

A Note for Parents & Guardians

Ten tests is a comprehensive program. Plan ***one test every 3–4 days*** *(or one per week) with study sessions between each test.*

How to help:

- *Tests 1–3 should be untimed — build understanding before adding pressure.*
- *After each test, review the answer explanations together. Ask: "Which topics were hardest? Let's study those before the next one."*
- *Use the Test Tracker to celebrate progress over time.*
- *For topic-specific help, pair this book with our* ***Grade 8 Math Study Guide*** *or* ***Grade 8 Workbook****.*

Grade 8 Formula Reference

Keep this page handy — you may use it during your practice tests!

Exponent Rules

$a^m \cdot a^n = a^{m+n}$ $(a^m)^n = a^{mn}$ $(ab)^n = a^n \cdot b^n$

$\frac{a^m}{a^n} = a^{m-n}$ $a^0 = 1$ $(a \neq 0)$ $a^{-n} = \frac{1}{a^n}$

Lines & Linear Equations

Slope: $m = \frac{y_2 - y_1}{x_2 - x_1} = \frac{\text{rise}}{\text{run}}$

m = slope b = y-intercept

Slope-intercept: $y = mx + b$

Parallel lines: same slope

Proportional: $y = mx$

Proportional: passes through origin

Scientific Notation

$a \times 10^n$ where $1 \leq |a| < 10$

Multiply: add exponents

Divide: subtract exponents

Roots & Number Sense

Perfect squares: 1, 4, 9, 16, 25, 36, 49, 64, 81, 100, 121, 144

Perfect cubes: 1, 8, 27, 64, 125 $\sqrt{2} \approx 1.414$ $\sqrt{3} \approx 1.732$ $\pi \approx 3.14159$

Pythagorean Theorem & Distance

$a^2 + b^2 = c^2$ c = hypotenuse (longest side of a right triangle)

Distance: $d = \sqrt{(x_2 - x_1)^2 + (y_2 - y_1)^2}$

Volume Formulas

Cylinder $V = \pi r^2 h$ **Cone** $V = \frac{1}{3}\pi r^2 h$ **Sphere** $V = \frac{4}{3}\pi r^3$

Angle Relationships

Triangle angle sum: $180°$

Exterior angle = *sum of two remote interior angles*

Parallel lines + transversal: *Alternate interior angles are equal* • *Co-interior angles sum to* $180°$

Functions

Each input → exactly one output

Vertical line test: *if any vertical line hits graph more than once ⇒ not a function*

Linear: *constant rate of change* $(y = mx + b)$

Nonlinear: *rate of change varies*

Transformations

Translation: *slide*

Reflection: *flip*

Rotation: *turn*

Dilation: *resize*

Congruent = same shape & size

Similar = same shape, proportional size

Tip: *Bookmark this page! Review it before each test so these formulas become second nature.*

Multiplication Table

You may use this table during your practice tests!

×	1	2	3	4	5	6	7	8	9	10	11	12
1	1	2	3	4	5	6	7	8	9	10	11	12
2	2	4	6	8	10	12	14	16	18	20	22	24
3	3	6	9	12	15	18	21	24	27	30	33	36
4	4	8	12	16	20	24	28	32	36	40	44	48
5	5	10	15	20	25	30	35	40	45	50	55	60
6	6	12	18	24	30	36	42	48	54	60	66	72
7	7	14	21	28	35	42	49	56	63	70	77	84
8	8	16	24	32	40	48	56	64	72	80	88	96
9	9	18	27	36	45	54	63	72	81	90	99	108
10	10	20	30	40	50	60	70	80	90	100	110	120
11	11	22	33	44	55	66	77	88	99	110	121	132
12	12	24	36	48	60	72	84	96	108	120	132	144

How to Use This Table

To find **4** × **7**:

1. Find **4** in the left column (blue).
2. Find **7** in the top row (blue).
3. Follow the row and column until they meet: the answer is **28**!

Tip: *You can also use this table for division! If you know* $28 \div 4 = ?$, *find* 28 *in the* 4*'s row. The column header gives you the answer:* **7**!

My Test Tracker

Record every test and watch your scores improve

Name: ______________________ **Start Date:** ______________

GETTING STARTED (Tests 1–3)

Test 1 — Untimed

Date: __________ Score: ______/______ %: ______ Topics to review: ______________

Test 2 — Untimed

Date: __________ Score: ______/______ %: ______ Topics to review: ______________

Test 3 — Soft Timer

Date: __________ Score: ______/______ %: ______ Topics to review: ______________

BUILDING SKILLS (Tests 4–7)

Test 4 — Timed (70 min)

Date: __________ Score: ______/______ %: ______ Focus area: ______________

Test 5 — Timed (70 min)

Date: __________ Score: ______/______ %: ______ Focus area: ______________

Test 6 — Timed (70 min)

Date: __________ Score: ______/______ %: ______ Focus area: ______________

Test 7 — Timed (70 min)

Date: __________ Score: ______/______ %: ______ Focus area: ______________

TEST-DAY READY (Tests 8–10)

Test 8 — Full Test Conditions

Date: ____________ Score: _______/_______ %: _______ Growth since Test 1: ____________________

Test 9 — Full Test Conditions

Date: ____________ Score: _______/_______ %: _______ Growth since Test 1: ____________________

Test 10 — Full Test Conditions

Date: ____________ Score: _______/_______ %: _______ Growth since Test 1: ____________________

Score Progress

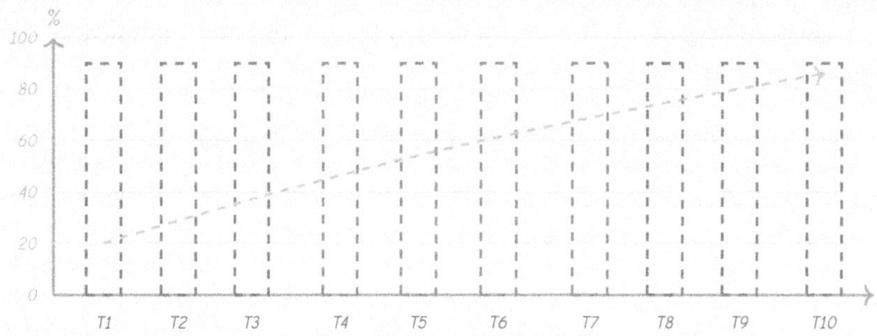

Shade each bar after every test. Watch your improvement!

Final Reflection

The most important thing I learned: ____

The topic where I improved the most: ____

My advice for other students: ____

YOUR ONLINE COMPANION

Continue Learning at ViewMath Academy!

For Parents, Teachers & Students

*Great job on the practice tests! Want to keep improving? ViewMath Academy is your **free online companion** to this book.*

- ***Score Analyzer** — Enter your answers and instantly see which topics need more practice*
- ***Interactive Lessons** — Review the concepts behind each question with clear explanations*
- ***Adaptive Quizzes** — Practice your weak topics with questions that match your level*
- ***Progress Tracking** — See your mastery grow across all Grade 8 math topics*
- ***Personalized Dashboard** — A learning plan tailored just for you*

Scan to visit ViewMath Academy

ViewMath.com/OK-Grade8

Free to use • No downloads required • Works on any device

Table of Contents

Here's what we'll explore together!

Let's learn and have fun!

PRACTICE TEST

1

Practice Test 1

30 Questions

Before You Start

- **Read each question carefully** before choosing your answer.
- **Show your work** on scratch paper when you need to.
- **Skip hard questions** and come back to them later.
- **Check your answers** when you're done.
- **Take your time** — there's no rush!

You've Got This!

Do your best and show what you know!

1. *True or false: The number* $\frac{22}{7}$ *is equal to* π.

Your Answer

2. *The table below shows four repeating decimals and the power of* 10 *a student used to convert each. Which one uses the WRONG power of* 10*?*

Decimal	Multiply by
$0.\overline{7}$	10
$0.\overline{45}$	100
$0.\overline{123}$	100
$0.\overline{8}$	10

(A) $0.\overline{7}$

(B) $0.\overline{45}$

(C) $0.\overline{123}$

(D) $0.\overline{8}$

3. *The table below shows a student's work to approximate* $\sqrt{14}$. *What should go in the blank?*

Guess	Square	Compare to 14
3	9	too small
4	16	too big
3.7	13.69	too small
3.8	14.44	too big
?	?	closest

(A) 3.74, *because* $3.74^2 = 13.9876$

(B) 3.72, *because* $3.72^2 = 13.8384$

(C) 3.80, *because* $3.8^2 = 14.44$

(D) 3.75, *because* $3.75^2 = 14.0625$

4. *What is the best estimate for* $5 - \sqrt{7}$?

(A) 1.4

(B) 2.4

(C) 3.0

(D) 0.4

5. *Look at the receipt below. The sales tax rate is 8%. What should the total be?*

Item	**Price**
T-shirt	$20.00
Jeans	$45.00
Subtotal	$65.00
Tax (8%)	?
Total	?

(A) Tax = $5.00; Total = $70.00

(B) Tax = $5.20; Total = $70.20

(C) Tax = $6.50; Total = $71.50

(D) Tax = $8.00; Total = $73.00

6. *Evaluate* $(-1)^{15}$.

Your Answer:

7. The table shows the approximate diameters of objects in our solar system. Which object has the greatest diameter?

Object	Diameter (km)
Moon	3.5×10^3
Earth	1.3×10^4
Jupiter	1.4×10^5
Sun	1.4×10^6

(A) Moon
(B) Earth
(C) Jupiter
(D) Sun

8. A swimming pool holds 8×10^4 liters of water. If a hose fills the pool at 2×10^2 liters per minute, how many minutes does it take to fill the pool?

(A) 4×10^6
(B) 4×10^2
(C) 6×10^2
(D) 1.6×10^7

9. Find the slope through $(3, -1)$ and $(7, 11)$.

Your Answer:

10. Find the equation of the line through $(2, 1)$ and $(6, 9)$.

Your Answer:

11. *The graph below shows a relation. Is it a function?*

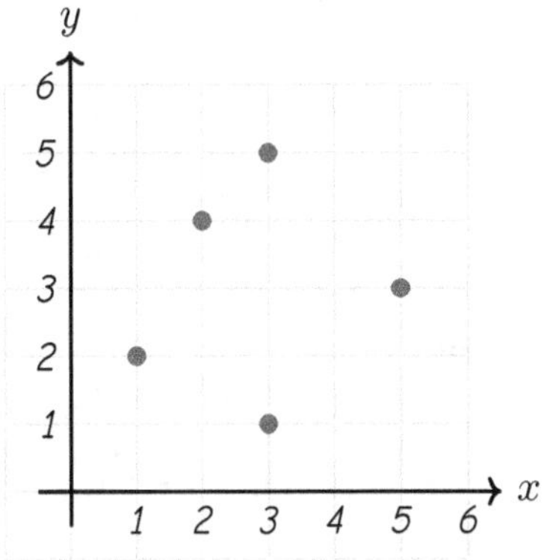

(A) Yes, because all the points are in the first quadrant.

(B) Yes, because no two points have the same y-value.

(C) No, because there are two points at $x = 3$.

(D) No, because the points do not form a line.

12. *The graph of h is shown below. For what value of x does $h(x) = 6$?*

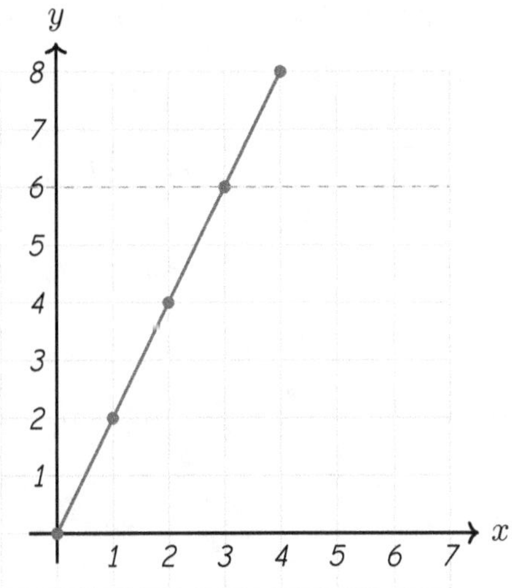

Your Answer:

13. Function M has slope 4 and y-intercept 10. Function N has slope 4 and y-intercept 3. Are their graphs parallel? Write Yes or No.

Your Answer:

14. Which equation represents a linear function?

(A) $y = x^2 + 3$

(B) $y = \frac{1}{x}$

(C) $y = 4x - 7$

(D) $y = 2^x$

15. The table shows $(2, 9)$ and $(6, 25)$. Find the slope of the line.

Your Answer:

16. A student walks to a friend's house, waits for the friend, then they walk together to school. What does the distance-from-home graph look like?

(A) Increasing, constant, increasing

(B) Increasing, decreasing, increasing

(C) Constant, increasing, constant

(D) Decreasing, constant, decreasing

17. Name the three rigid transformations.

Your Answer:

18. $\triangle PQR \cong \triangle STU$. If $\angle P = 72°$ and $\angle Q = 53°$, what is $\angle U$?

Your Answer:

19. *A triangle with vertices* $(0,0)$, $(6,0)$, *and* $(6,8)$ *is dilated by factor* $\frac{1}{2}$ *from the origin. What is the perimeter of the image?*

(A) 24
(B) 12
(C) 6
(D) 48

20. *A dilation centered at the origin maps* $(4,-2)$ *to* $(12,-6)$. *What is the scale factor?*

(A) 2
(B) 3
(C) 4
(D) 6

21. *Which angles are equal when parallel lines are cut by a transversal?*

(A) *Co-interior angles*
(B) *Supplementary angles*
(C) *Corresponding angles*
(D) *Adjacent angles*

22. *A right triangle has legs* 5 *and* 5. *What is the hypotenuse?*

(A) 10
(B) $\sqrt{50}$
(C) 25
(D) $\sqrt{10}$

23. *What is the distance between* $(-2,1)$ *and* $(4,9)$?

(A) 10
(B) $\sqrt{10}$
(C) 14
(D) 8

24. *A sphere's volume depends only on its:*

(A) *Height*

(B) *Diameter and height*

(C) *Radius*

(D) *Base area and height*

25. *A cylinder has a surface area of 200π cm^2 and a radius of 5 cm. What is the height?*

(A) 10 *cm*

(B) 15 *cm*

(C) 20 *cm*

(D) 8 *cm*

26. *What type of association does this scatter plot show?*

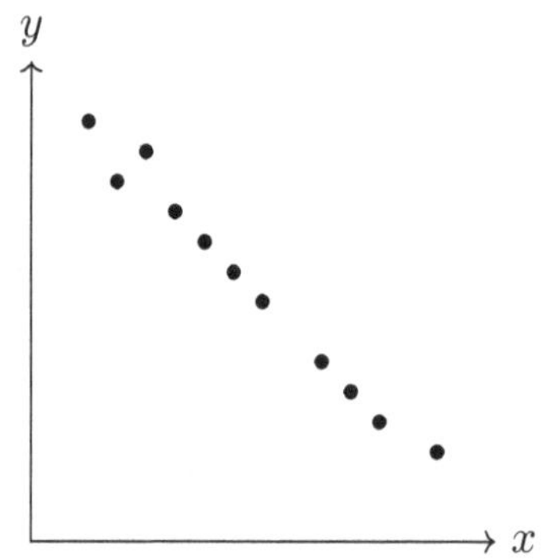

(A) *Positive linear*

(B) *Negative linear*

(C) *No association*

(D) *Nonlinear*

27. *A scatter plot has 10 data points. One outlier is far above the trend. When drawing the line of best fit, you should:*

(A) *draw the line through the outlier*

(B) *ignore the outlier and fit the line to the other 9 points*

(C) *not draw a line at all*

(D) *draw the line only through the outlier*

28. A trend line gives $y = -3x + 45$. Data was collected for $x = 1$ to $x = 10$. Predict y when $x = 5$ and when $x = 30$. State which prediction is more reliable and why.

Your Answer:

29. Using the table above, what percentage of 7th graders ride a bike?

(A) 55%

(B) 45%

(C) 62.5%

(D) 37.5%

30. A deck has 52 cards. Two cards are drawn without replacement. Find $P(\text{both aces})$.

Your Answer:

End of Practice Test 1

Great job finishing the test!

My Score

I got __________ out of 30 questions right.

*Check your answers in the **Answer Key** at the back of the book.*

Review any questions you missed. That's how we learn!

Check Your Score Online!

*Visit **ViewMath Academy** to enter your answers and see which topics you need to review. You can also explore lessons, take quizzes, track your scores, and save your progress!*

viewmath.com/score/8.1.OK.16

Or go to **viewmath.com/score** *and enter code:* **8.1.OK.16**

PRACTICE TEST

2

Practice Test 2

30 Questions

Before You Start

- **Read each question carefully** before choosing your answer.
- **Show your work** on scratch paper when you need to.
- **Skip hard questions** and come back to them later
- **Check your answers** when you're done.
- **Take your time** — there's no rush!

You've Got This!

Do your best and show what you know!

1. The number line below shows two points, P and Q.

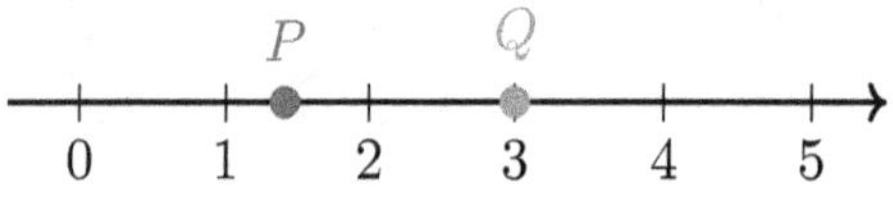

Point P represents $\sqrt{2}$ and point Q represents 3. Which statement is true?

(A) Both P and Q are rational.

(B) Both P and Q are irrational.

(C) P is irrational and Q is rational.

(D) P is rational and Q is irrational.

2. The flowchart below shows the conversion steps for a repeating decimal. Fill in the missing values to convert $0.\overline{54}$ to a fraction.

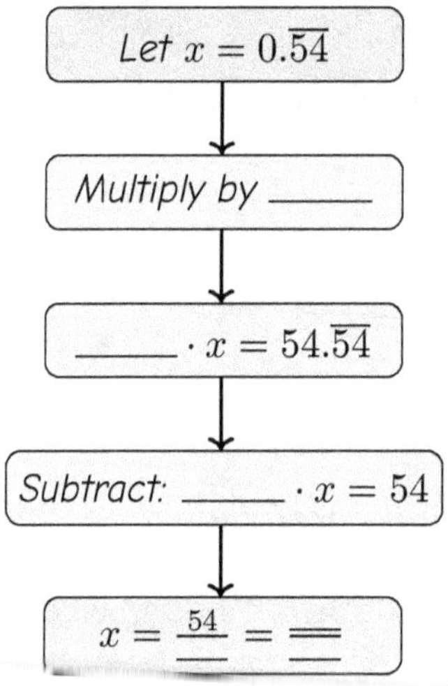

Your Answer:

3. The squares below have the given areas. Estimate the side length of each square to one decimal place.

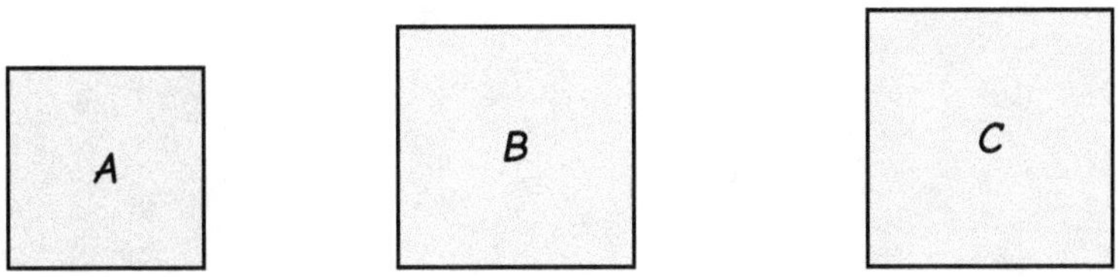

Area $= 18$ sq units Area $= 40$ sq units Area $= 72$ sq units

Your Answer:

4. What is the best estimate for $\sqrt{5} + \sqrt{3}$?

(A) $\sqrt{8} \approx 2.8$

(B) 3.0

(C) 4.0

(D) 8.0

5. An investment of \$800 earns \$96 in simple interest over 4 years. What is the annual interest rate?

(A) 2%

(B) 3%

(C) 4%

(D) 12%

6. Which of the following is equal to $\frac{8^6}{8^6}$?

(A) 8^{36}

(B) 8^0

(C) 0

(D) 8

7. Which of the following is 0.00072 written in scientific notation?

(A) 7.2×10^{-4}

(B) 7.2×10^{4}

(C) 72×10^{-5}

(D) 0.72×10^{-3}

8. Compute $(8 \times 10^{-4})(5 \times 10^{-3})$. Write your answer in scientific notation.

Your Answer:

9. A plant grows from 4 cm to 16 cm in 6 weeks. What is the rate of change?

(A) 1 cm/week
(B) 2 cm/week
(C) 3 cm/week
(D) 4 cm/week

10. What is the slope of $y = -\frac{5}{2}x + 8$?

Your Answer:

11. Given $\{(-2,5),(-1,3),(0,1),(1,-1)\}$, what is the output when the input is 0?

(A) -1
(B) 0
(C) 1
(D) 5

12. A function table shows $f(1) = 4$, $f(2) = 7$, $f(3) = 10$. What is $f(2) - f(1)$?

Your Answer:

13. The tables below show two functions. Find the rate of change and initial value of each. Which function will have a greater value at $x = 10$?

Function A

x	y
0	20
1	23
2	26
3	29

Function B

x	y
0	5
1	10
2	15
3	20

Your Answer:

14. Give an example of a nonlinear equation.

Your Answer:

15. A line passes through $(1, 6)$ and $(4, 18)$. What is the slope?

(A) 3
(B) 4
(C) 6
(D) 12

16. A bathtub fills up, then someone soaks, then the water drains. Which describes the water level graph?

(A) Decreasing, constant, increasing
(B) Increasing, decreasing, constant
(C) Increasing, constant, decreasing
(D) Constant, increasing, decreasing

17. A figure is reflected over the y-axis. What happens to point $(7, -2)$?

(A) $(7, 2)$
(B) $(-7, 2)$
(C) $(-7, -2)$
(D) $(-2, 7)$

18. Triangle $ABC \cong$ Triangle DEF. If $AB = 8$ cm, what is DE?

(A) 4 cm
(B) 8 cm
(C) 16 cm
(D) Cannot be determined

19. Translate the point $(4, -6)$ by $(-3, 8)$. What is the image?

Your Answer:

20. A rectangle is 3 cm by 5 cm. Another rectangle is 6 cm by 9 cm. Are they similar?

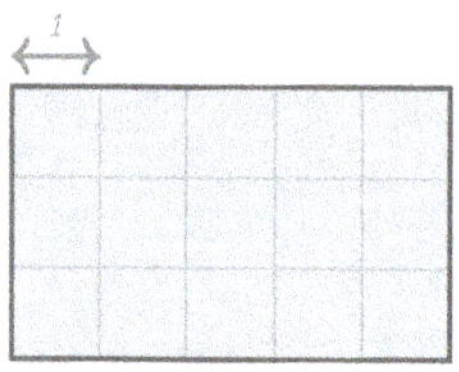

(A) Yes, with scale factor 2
(B) Yes, with scale factor 3
(C) No, the sides are not proportional.
(D) Yes, with scale factor $\frac{3}{5}$

21. Two angles of a triangle are $37°$ and $53°$. What is the exterior angle adjacent to the third angle?

(A) $90°$
(B) $127°$
(C) $53°$
(D) $143°$

22. Is a triangle with sides 5, 10, 12 a right triangle?

Your Answer:

23. *What is the distance between $(0,0)$ and $(5,0)$?*

(A) 0

(B) 5

(C) 25

(D) $\sqrt{5}$

24. *A cone has radius 5 cm and height 12 cm. What is its volume in terms of π?*

Your Answer:

25. *Find the total surface area of the cylinder. (Use $\pi \approx 3.14$)*

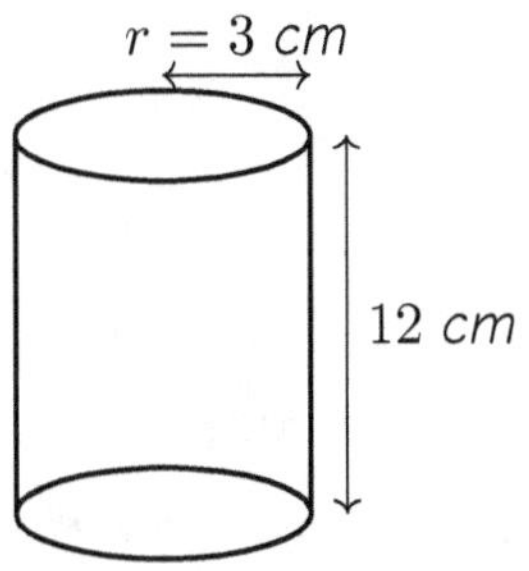

(A) 282.6 cm^2

(B) 226.08 cm^2

(C) 339.12 cm^2

(D) 113.04 cm^2

26. Describe the association shown in this scatter plot. Mention direction, shape, and any unusual features.

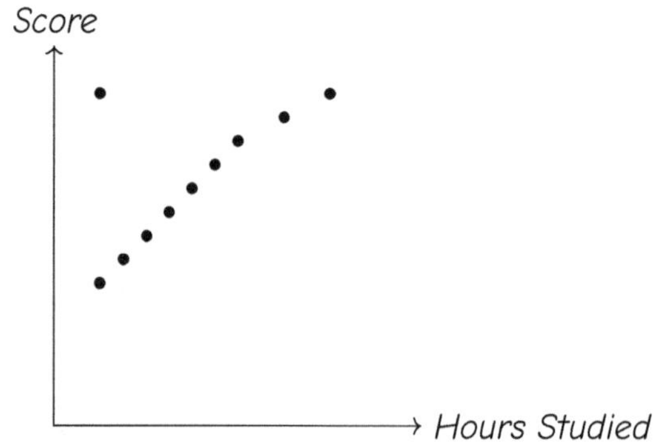

Your Answer

27. What is a line of best fit?

(A) A line that passes through every data point

(B) A line that connects the first and last data points

(C) A straight line that comes close to most of the data points

(D) A vertical line through the middle of the scatter plot

28. Which statement about slope in a linear model is true?

(A) A positive slope means y decreases as x increases.

(B) A negative slope means y increases as x increases.

(C) A slope of zero means y stays constant.

(D) The slope is always positive in real data.

29. A segmented (stacked) bar chart shows the proportion of students choosing band or chorus by gender. The boys' bar is 60% band and 40% chorus. The girls' bar is 45% band and 55% chorus. What can you conclude?

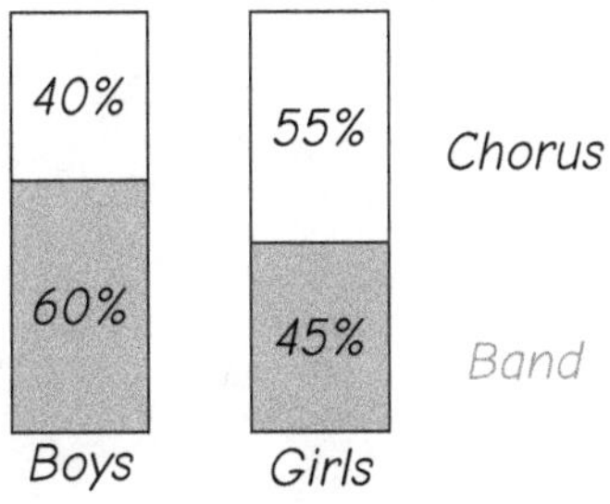

Ⓐ There is no association between gender and music choice.

Ⓑ Boys are more likely to choose band; girls more likely to choose chorus.

Ⓒ Girls and boys prefer band equally.

Ⓓ Chorus is more popular overall.

30. A tree diagram shows flipping a coin and rolling a die $(1–6)$. How many total outcomes are in the sample space?

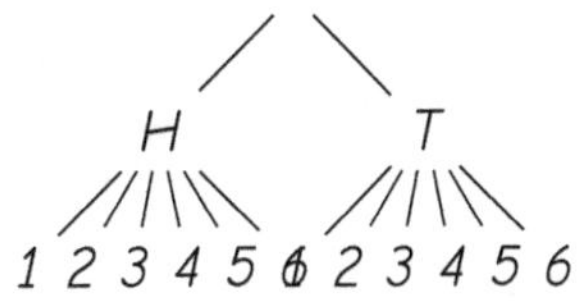

Ⓐ 0

Ⓑ 8

Ⓒ 12

Ⓓ 36

★ End of Practice Test 2 ★

Great job finishing the test!

My Score

I got __________ out of 30 questions right.

*Check your answers in the **Answer Key** at the back of the book.*

Review any questions you missed. That's how we learn!

Check Your Score Online!

*Visit **ViewMath Academy** to enter your answers and see which topics you need to review. You can also explore lessons, take quizzes, track your scores, and save your progress!*

viewmath.com/score/8.1.OK.17

*Or go to **viewmath.com/score** and enter code: **8.1.OK.17***

PRACTICE TEST

3

Practice Test 3

30 Questions

Before You Start

- **Read each question carefully** before choosing your answer.
- **Show your work** on scratch paper when you need to.
- **Skip hard questions** and come back to them later.
- **Check your answers** when you're done.
- **Take your time** — there's no rush!

You've Got This!

Do your best and show what you know!

1. *Which number below can be written as a fraction $\frac{a}{b}$ where a and b are integers and $b \neq 0$?*

Ⓐ $\sqrt{3}$

Ⓑ π

Ⓒ $0.\overline{81}$

Ⓓ $\sqrt{11}$

2. *What is $0.\overline{27}$ as a fraction in simplest form?*

Ⓐ $\frac{27}{100}$

Ⓑ $\frac{27}{99}$

Ⓒ $\frac{3}{11}$

Ⓓ $\frac{9}{33}$

3. *A student estimated $\sqrt{60} \approx 8$. Is this estimate reasonable?*

Ⓐ *Yes, because $8^2 = 64$ is close to 60.*

Ⓑ *No, because $8^2 = 64$, which is too far from 60. A better estimate is about 7.7.*

Ⓒ *No, $\sqrt{60} = 30$.*

Ⓓ *Yes, because $60 \div 8 \approx 8$.*

4. *Which is larger: $4\sqrt{2}$ or $\sqrt{30}$?*

Ⓐ $4\sqrt{2}$, *because* $4 > \sqrt{30}$

Ⓑ $\sqrt{30}$, *because* $30 > 2$

Ⓒ $4\sqrt{2}$, *because* $4\sqrt{2} \approx 5.66 > 5.48 \approx \sqrt{30}$

Ⓓ *They are equal.*

5. *A shirt originally costs \$35. It is marked down 20%, then an 8% sales tax is added. What is the final price?*

Ⓐ \$30.24

Ⓑ \$28.00

Ⓒ \$32.76

Ⓓ \$31.08

6. Simplify $\frac{5^7}{5^3}$.

(A) 5^{10}
(B) 5^4
(C) 1^4
(D) 5^{21}

7. Earth is about 1.5×10^8 km from the Sun. Mars is about 2.3×10^8 km from the Sun. About how much farther from the Sun is Mars than Earth?

(A) 0.8×10^8 km
(B) 8×10^7 km
(C) 3.8×10^8 km
(D) Both A and B

8. What is $5.2 \times 10^6 + 3.8 \times 10^6$?

(A) 9×10^6
(B) 9×10^{12}
(C) 19.76×10^6
(D) 9×10^{36}

9. A car's odometer reads 120 miles at 2:00 PM and 270 miles at 5:00 PM. What is the car's speed (rate of change)?

(A) 45 mph
(B) 50 mph
(C) 60 mph
(D) 90 mph

10. A taxi charges a \$3 base fare plus \$2 per mile. Which equation models the total cost y after x miles?

(A) $y = 3x + 2$
(B) $y = 2x + 3$
(C) $y = 5x$
(D) $y = 2x - 3$

11. In the ordered pair (x, y), x is called the:

Ⓐ output
Ⓑ range
Ⓒ input
Ⓓ function

12. If $f(x) = 4x + 1$, find the value of $f(3) + f(1)$.

Your Answer:

13. Function A starts at $\$50$ and increases by $\$12$ per week. Function B: $y = 15x + 30$. Which earns more per week?

Ⓐ Function A
Ⓑ Function B
Ⓒ They earn the same per week.
Ⓓ Cannot be determined.

14. Is $y = x^3 - 2$ linear or nonlinear?

Your Answer:

15. A phone plan costs $\$35$ per month with no activation fee. Which equation models the total cost?

Ⓐ $y = 35$
Ⓑ $y = 35x$
Ⓒ $y = 35x + 35$
Ⓓ $y = x + 35$

16. A roller coaster goes up, down sharply, up again, then levels off. How many sections does the graph have?

Ⓐ 2
Ⓑ 3
Ⓒ 4
Ⓓ 5

17. The figure below shows a point P and its image P' after a transformation. Which transformation was applied?

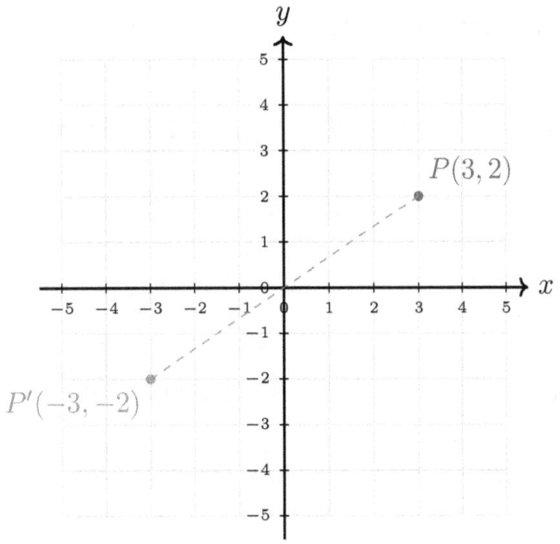

Ⓐ Reflection over the x-axis

Ⓑ Reflection over the y-axis

Ⓒ Rotation $90°$ counterclockwise

Ⓓ Rotation $180°$ around the origin

18. Which transformation would NOT preserve congruence?

Ⓐ Reflection over the x-axis

Ⓑ Translation 2 units up

Ⓒ Dilation by a factor of 1.5

Ⓓ Rotation of $270°$

19. A point $(-1, 6)$ is rotated $90°$ clockwise around the origin. The $90°$ clockwise rule is $(x, y) \to (y, -x)$. What is the image?

Ⓐ $(6, 1)$

Ⓑ $(-6, -1)$

Ⓒ $(-6, 1)$

Ⓓ $(1, -6)$

20. *Are all squares similar to each other?*

Ⓐ No, because they can have different side lengths.

Ⓑ No, because they can have different angles.

Ⓒ Yes, because all squares have equal angles and proportional sides.

Ⓓ Yes, but only if they have the same perimeter.

21. *An exterior angle of a triangle is always:*

Ⓐ Less than any interior angle

Ⓑ Equal to the adjacent interior angle

Ⓒ Greater than either non-adjacent interior angle

Ⓓ Equal to 180°

22. *Find the missing side x in the right triangle below.*

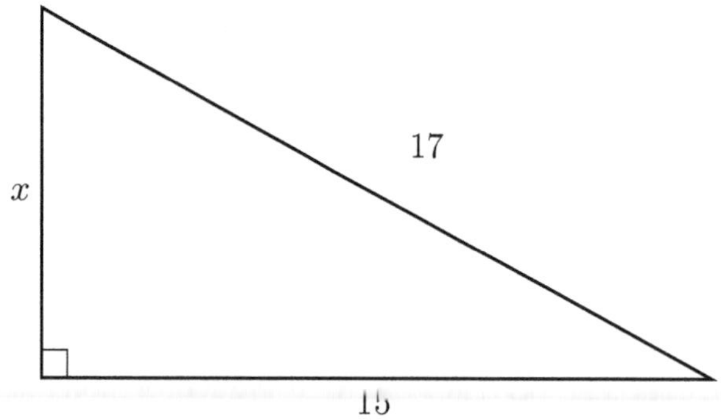

Your Answer:

23. *Two corners of a square are at $(0,0)$ and $(0,6)$. What is the length of the diagonal?*

Ⓐ 6

Ⓑ 12

Ⓒ $6\sqrt{2}$

Ⓓ 36

24. A basketball has a diameter of 24 cm. What is its volume? Use $\pi \approx 3.14$

(A) 7,234.6 cm^3

(B) 2,143.6 cm^3

(C) 904.3 cm^3

(D) 18,086.4 cm^3

25. A square pyramid has total surface area 360 in^2 and a base side of 12 in. What is the slant height?

(A) 7 in

(B) 9 in

(C) 10 in

(D) 14 in

26. Can a scatter plot show a positive association that is nonlinear? Give an example.

Your Answer:

27. A scatter plot shows a strong nonlinear (curved) pattern. Should you draw a straight line of best fit?

(A) Yes — always draw a straight line.

(B) No — a straight line does not fit curved data well.

(C) Yes — but only if there are outliers.

(D) No — you should never draw any line.

28. Using the model $y = -4x + 100$, predict y when $x = 10$.

(A) 140

(B) 60

(C) 96

(D) 40

29.

	Dog	Cat	Total
Boys	15	10	25
Girls	12	13	25
Total	27	23	50

How many girls prefer dogs?

(A) 10

(B) 12

(C) 13

(D) 15

30. A fair die is rolled twice. What is $P(\text{both rolls are } 6)$?

(A) $\frac{1}{6}$

(B) $\frac{1}{12}$

(C) $\frac{1}{36}$

(D) $\frac{2}{6}$

★ End of Practice Test 3 ★

Great job finishing the test!

My Score

I got __________ out of 30 questions right.

*Check your answers in the **Answer Key** at the back of the book.*

Review any questions you missed. That's how we learn!

Check Your Score Online!

*Visit **ViewMath Academy** to enter your answers and see which topics you need to review. You can also explore lessons, take quizzes, track your scores, and save your progress!*

viewmath.com/score/8.1.OK.18

*Or go to **viewmath.com/score** and enter code: **8.1.OK.18***

PRACTICE TEST

4

Practice Test 4

30 Questions

Before You Start

- **Read each question carefully** before choosing your answer.
- **Show your work** on scratch paper when you need to.
- **Skip hard questions** and come back to them later.
- **Check your answers** when you're done.
- **Take your time** — there's no rush!

You've Got This!

Do your best and show what you know!

1. Which of the following numbers is irrational?

(A) $\frac{5}{8}$

(B) $0.\overline{6}$

(C) $\sqrt{9}$

(D) $\sqrt{7}$

2. Write $0.5\overline{8}$ as a fraction in simplest form.

Your Answer:

3. Which inequality is true?

(A) $\sqrt{15} < 3$

(B) $\sqrt{15} > 4$

(C) $\sqrt{15} > 3$

(D) $\sqrt{15} = 4$

4. Compute $\sqrt{2} \times \sqrt{18}$ exactly, without a calculator.

Your Answer:

5. What is the simple interest on $\$600$ at 5% per year for 3 years?

(A) $\$30$

(B) $\$90$

(C) $\$900$

(D) $\$9$

6. What is the value of $(-2)^0$?

(A) -2

(B) 0

(C) 1

(D) -1

7. Order these numbers from greatest to least: 7.1×10^{-2}, 7.1×10^{2}, 7.1×10^{0}

(A) $7.1 \times 10^{-2},\ 7.1 \times 10^{0},\ 7.1 \times 10^{2}$
(B) $7.1 \times 10^{2},\ 7.1 \times 10^{-2},\ 7.1 \times 10^{0}$
(C) $7.1 \times 10^{2},\ 7.1 \times 10^{0},\ 7.1 \times 10^{-2}$
(D) $7.1 \times 10^{0},\ 7.1 \times 10^{2},\ 7.1 \times 10^{-2}$

8. Add $2.8 \times 10^{5} + 7.2 \times 10^{5}$. Write your answer in scientific notation.

Your Answer:

9. What is the slope through $(-4, -1)$ and $(2, 5)$?

(A) -1
(B) $\frac{2}{3}$
(C) 1
(D) $\frac{3}{2}$

10. A line passes through $(0, 4)$ and $(2, 10)$. What is its equation?

(A) $y = 2x + 4$
(B) $y = 3x + 4$
(C) $y = 4x + 2$
(D) $y = 5x + 4$

11. Does the equation $y = -4x + 9$ represent a function?

(A) No, because the slope is negative.
(B) No, because y can be negative.
(C) Yes, because each value of x gives exactly one value of y.
(D) Yes, because the equation has two variables.

12. If $f(x) = 8x - 6$ and $f(x) = 42$, what is x?

Your Answer:

13. The table shows Function P. Function Q is $y = 2x + 9$. Which function has the greater initial value?

x	0	1	2	3
$P(x)$	5	8	11	14

(A) Function P

(B) Function Q

(C) They have the same initial value.

(D) Cannot be determined.

14. A table shows $(1, 1)$, $(2, 8)$, $(3, 27)$, $(4, 64)$. Is this linear or nonlinear? What pattern do you see?

Your Answer

15. The function $y = 50x + 200$ models a bank account balance after x weeks. What is the weekly deposit?

(A) $\$50$

(B) $\$100$

(C) $\$200$

(D) $\$250$

16. Which graph best matches this story: "Maria walks quickly to the store, waits in line, then walks slowly home"?

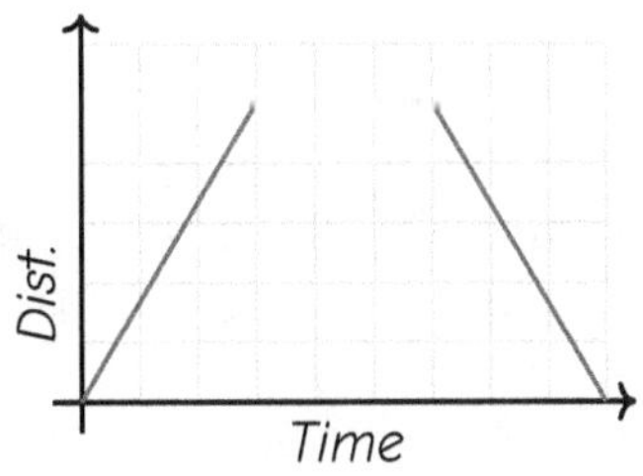

Graph A

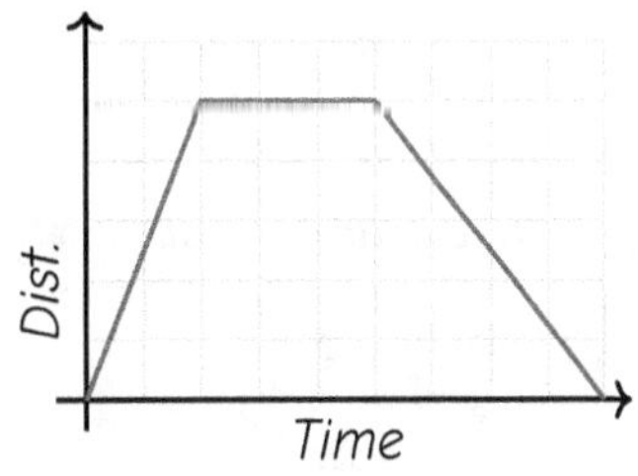

Graph B

(A) Graph A only

(B) Graph B only

(C) Both graphs match equally.

(D) Neither graph matches.

17. *Which transformation turns a figure around a fixed point?*

- (A) *Translation*
- (B) *Reflection*
- (C) *Rotation*
- (D) *Dilation*

18. *A student says two circles with radius* 5 *cm are always congruent. Is this correct?*

- (A) *No, circles cannot be congruent.*
- (B) *No, they need the same center too.*
- (C) *Yes, because they have the same radius.*
- (D) *Yes, but only if they are in the same position.*

19. *Point* (a, b) *is rotated* $180°$ *and then reflected over the* x*-axis. The final image is:*

- (A) (a, b)
- (B) $(-a, b)$
- (C) $(a, -b)$
- (D) $(-a, -b)$

20. *A dilation with* $k = 1$ *produces:*

- (A) *A figure twice as large*
- (B) *A congruent figure*
- (C) *A figure half as large*
- (D) *No figure at all*

21. *Two parallel lines are cut by a transversal. One angle measures* $72°$*. What is its alternate interior angle?*

- (A) $18°$
- (B) $72°$
- (C) $108°$
- (D) $288°$

22. A ladder 15 ft long leans against a wall. The base is 9 ft from the wall. How high up the wall does the ladder reach?

(A) 6 ft

(B) 12 ft

(C) $\sqrt{306}$ ft

(D) 24 ft

23. Points P and Q are shown on the grid. What is the distance between them?

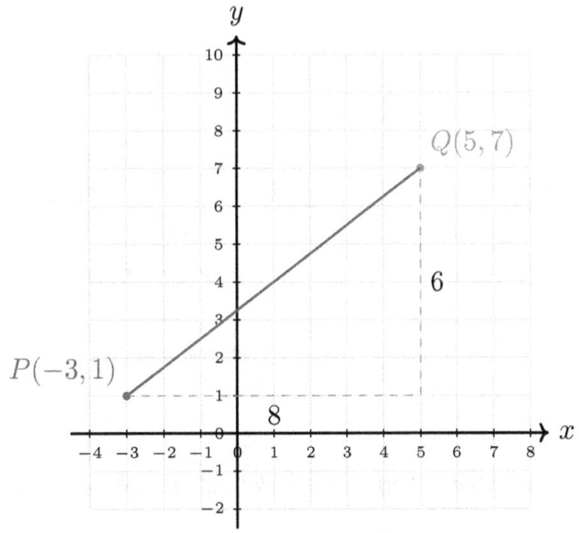

(A) 14

(B) $\sqrt{14}$

(C) 10

(D) $\sqrt{28}$

24. A cone has volume 100 cm^3. A cylinder with the same base and height has volume ___ cm^3.

Your Answer:

25. Find the lateral surface area of a cylinder with radius 7 cm and height 10 cm. (Use $\pi \approx 3.14$.)

(A) 439.6 cm^2

(B) 219.8 cm^2

(C) 307.72 cm^2

(D) 747.32 cm^2

26. *A scatter plot shows data points that form a curved (U-shaped) pattern. This is best described as:*

Ⓐ *positive linear association*

Ⓑ *negative linear association*

Ⓒ *nonlinear association*

Ⓓ *no association*

27. *A trend line has equation* $y = 1.5x + 2$. *Using this, what is* y *when* $x = 4$?

Ⓐ 6

Ⓑ 7

Ⓒ 8

Ⓓ 10

28. *Data was collected for* x*-values from* 1 *to* 10. *Using the model to predict* y *when* $x = 50$ *is an example of:*

Ⓐ *interpolation*

Ⓑ *extrapolation*

Ⓒ *correlation*

Ⓓ *association*

29. 80 *students were surveyed about playing sports and playing a musical instrument.*

	Plays sport	No sport	Total
Plays instrument	15	25	40
No instrument	30	10	40
Total	45	35	80

Is there an association between playing a sport and playing an instrument? Justify using conditional relative frequencies.

Your Answer:

30. *A jar has* 6 *green and* 4 *yellow balls. One ball is drawn and replaced. Then another is drawn.* P(*both green*) $=$?

Ⓐ $\frac{36}{100}$

Ⓑ $\frac{30}{90}$

Ⓒ $\frac{6}{10}$

Ⓓ $\frac{12}{20}$

★ End of Practice Test 4 ★

Great job finishing the test!

My Score

I got __________ out of 30 questions right.

*Check your answers in the **Answer Key** at the back of the book.*

Review any questions you missed. That's how we learn!

Check Your Score Online!

*Visit **ViewMath Academy** to enter your answers and see which topics you need to review. You can also explore lessons, take quizzes, track your scores, and save your progress!*

viewmath.com/score/8.1.OK.19

*Or go to **viewmath.com/score** and enter code: **8.1.OK.19***

PRACTICE TEST

5

Practice Test 5

30 Questions

Before You Start

- **Read each question carefully** before choosing your answer.
- **Show your work** on scratch paper when you need to.
- **Skip hard questions** and come back to them later.
- **Check your answers** when you're done.
- **Take your time** — there's no rush!

You've Got This!

Do your best and show what you know!

1. *A calculator shows* $\sqrt{144} = 12$. *Is* $\sqrt{144}$ *rational or irrational? Explain.*

Your Answer:

2. *To convert* $0.\overline{63}$ *to a fraction, which equation should you set up after letting* $x = 0.\overline{63}$?

(A) $10x - x = 63$

(B) $100x - x = 63$

(C) $100x - x = 6.3$

(D) $1000x - x = 63$

3. *Which inequality correctly describes the location of* $\sqrt{22}$?

(A) $4 < \sqrt{22} < 5$

(B) $5 < \sqrt{22} < 6$

(C) $10 < \sqrt{22} < 12$

(D) $\sqrt{22} = 11$

4. *The diagram below shows a rectangle with irrational side lengths. Which is the best estimate of the area?*

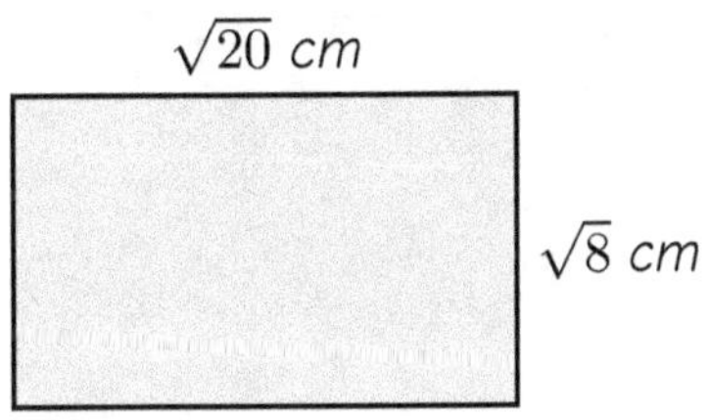

(A) $\sqrt{28} \approx 5.3$ cm^2

(B) $\sqrt{160} \approx 12.6$ cm^2

(C) 28 cm^2

(D) 160 cm^2

5. *A* $120 *pair of shoes is on sale for* 25% *off. What is the sale price?*

(A) $30

(B) $95

(C) $90

(D) $25

6. Simplify $\frac{12^8}{12^8}$.

Your Answer:

7. What is 6.1×10^3 written in standard form?

(A) 61,000
(B) 610
(C) 6,100
(D) 0.0061

8. A bacterium has a length of 2×10^{-6} m. If 1 micrometer (μm) $= 10^{-6}$ m, what is the bacterium's length in micrometers?

(A) 2×10^{-12} μm
(B) 0.002 μm
(C) 2 μm
(D) 2,000 μm

9. Find the slope through $\left(\frac{1}{2}, 3\right)$ and $\left(\frac{3}{2}, 7\right)$.

(A) 2
(B) 4
(C) $\frac{1}{4}$
(D) 8

10. A line passes through $(0, 0)$ and $(7, 21)$. Write its equation.

Your Answer:

11. Is the relation $\{(3, 8), (4, 8), (5, 8), (6, 8)\}$ a function? Write Yes or No.

Your Answer:

12. *If* $f(x) = -2x + 10$, *what is* $f(0)$?

(A) -2
(B) 0
(C) 8
(D) 10

13. *A table for Function C shows* $(0, 7)$, $(1, 11)$, $(2, 15)$. *What is the initial value?*

Your Answer:

14. *A student says* $y = 3x^2$ *is linear because the coefficient is* 3. *What is wrong with this reasoning?*

(A) The coefficient should be 1 for it to be linear.
(B) The variable x is raised to the 2nd power, making it nonlinear.
(C) The function has no y-intercept.
(D) Nothing is wrong; it is linear.

15. *A line passes through* $(2, 10)$ *and* $(5, 22)$. *What is the equation?*

(A) $y = 4x + 2$
(B) $y = 3x + 4$
(C) $y = 4x + 10$
(D) $y = 12x - 2$

16. *A graph is flat for* 4 *seconds on a speed-time graph. What does this mean?*

Your Answer:

17. *A rectangle has vertices at* $(1, 2)$, $(5, 2)$, $(5, 4)$, *and* $(1, 4)$. *After a translation of* 3 *units right and* 1 *unit up, what are the coordinates of the vertex that was at* $(1, 2)$?

(A) $(4, 3)$
(B) $(4, 1)$
(C) $(2, 5)$
(D) $(-2, 3)$

18. Figure B is obtained by reflecting Figure A over a line and then translating it 3 units right. Are the figures congruent?

(A) No, because two transformations were used.
(B) No, because the position changed.
(C) Yes, because both transformations are rigid.
(D) Yes, but only if the figures are triangles.

19. Which transformation maps (x, y) to $(-y, x)$?

(A) Reflection over the x-axis
(B) Reflection over the y-axis
(C) $90°$ counterclockwise rotation
(D) $180°$ rotation

20. If two figures are congruent, are they also similar?

(A) Yes, with scale factor $k = 1$.
(B) No, congruent and similar are different properties.
(C) Only if they are triangles.
(D) Only if they have the same orientation.

21. In a triangle, the angles are $(2x + 5)°$, $(3x)°$, and $(x + 25)°$. Find x.

Your Answer:

22. Is a triangle with sides 7, 24, 25 a right triangle?

(A) Yes, because $7 + 24 > 25$.
(B) Yes, because $7^2 + 24^2 = 25^2$.
(C) No, because $7^2 + 24^2 \neq 25^2$.
(D) No, because all sides must be equal.

23. A line segment has endpoints $(2, 3)$ and $(10, 9)$. What is its length?

(A) 10
(B) $\sqrt{10}$
(C) 14
(D) 8

24. What is the volume of a cylinder with radius 4 cm and height 10 cm? Use $\pi \approx 3.14$

(A) $125.6\ cm^3$

(B) $502.4\ cm^3$

(C) $160\ cm^3$

(D) $251.2\ cm^3$

25. Which formula gives the lateral surface area of a right cylinder with radius r and height h?

(A) $\pi r^2 h$

(B) $2\pi rh$

(C) $2\pi r^2 + 2\pi rh$

(D) $\pi r(r+h)$

26. Which variable typically goes on the x-axis in a scatter plot?

(A) The dependent variable

(B) The response variable

(C) The independent (explanatory) variable

(D) The variable with the largest values

27. A trend line passes through $(2, 10)$ and $(8, 4)$. Find the slope and y-intercept.

Your Answer:

28. Using the model $y = 4x + 7$, for what value of x is $y = 35$?

(A) 7

(B) 6

(C) 8

(D) 10

29. Complete the two-way table and find the relative frequencies for each cell (as fractions of the total).

	Homework Yes	Homework No	Total
Grade A	24	6	
Not Grade A	16	14	
Total			

Your Answer:

30. $P(A) = 0.6$, $P(B) = 0.4$. If A and B are independent, $P(A \text{ and } B) =$?

(A) 1.0

(B) 0.24

(C) 0.20

(D) 0.10

End of Practice Test 5

Great job finishing the test!

My Score

I got __________ out of 30 questions right.

Check your answers in the ***Answer Key*** *at the back of the book.*

Review any questions you missed. That's how we learn!

Check Your Score Online!

Visit ***ViewMath Academy*** *to enter your answers and see which topics you need to review. You can also explore lessons, take quizzes, track your scores, and save your progress!*

viewmath.com/score/8.1.OK.20

Or go to **viewmath.com/score** *and enter code:* **8.1.OK.20**

PRACTICE TEST

6

Practice Test 6

30 Questions

Before You Start

- **Read each question carefully** before choosing your answer.
- **Show your work** on scratch paper when you need to.
- **Skip hard questions** and come back to them later.
- **Check your answers** when you're done.
- **Take your time** — there's no rush!

You've Got This!

Do your best and show what you know!

1. Is $\frac{5}{6}$ rational or irrational? Explain.

Your Answer:

2. When converting $0.\overline{123}$ to a fraction, by what power of 10 should you multiply both sides?

(A) 10
(B) 100
(C) 1000
(D) 10000

3. Between which two consecutive integers does $\sqrt{110}$ lie?

(A) 9 and 10
(B) 10 and 11
(C) 11 and 12
(D) 54 and 56

4. A student says $\sqrt{4}+\sqrt{9}=\sqrt{13}$. Is this correct?

(A) Yes, because you add the numbers under the square roots.
(B) No, $\sqrt{4}+\sqrt{9}=2+3=5$, but $\sqrt{13}\approx 3.6$.
(C) Yes, because $4+9=13$.
(D) No, $\sqrt{4}+\sqrt{9}=\sqrt{36}=6$.

5. Jake borrows $2,000 at 6% simple interest for 2 years. How much total does he pay back?

(A) $2,120
(B) $2,240
(C) $2,600
(D) $2,012

6. Which expression is equivalent to 4^{-3}?

(A) -64
(B) -12
(C) $\frac{1}{12}$
(D) $\frac{1}{64}$

7. The U.S. population is approximately 3.3×10^8. The world population is approximately 8×10^9. About how many times larger is the world population than the U.S. population?

(A) About 2.4 times
(B) About 24 times
(C) About 240 times
(D) About 2,400 times

8. Which expression equals $870{,}000 + 4.3 \times 10^4$?

(A) 9.13×10^5
(B) 8.743×10^5
(C) 4.387×10^5
(D) 8.7×10^5

9. A line passes through $(2, 7)$ and $(5, 1)$. What is the slope?

(A) -2
(B) 2
(C) -3
(D) 3

10. A line passes through $(1, 2)$ and $(4, 11)$. What is the equation in slope-intercept form?

(A) $y = 3x - 1$
(B) $y = 3x + 1$
(C) $y = 3x + 2$
(D) $y = 3x - 2$

11. Which of these equations represents a function?

(A) $x = 5$
(B) $y = 7$
(C) $x^2 + y^2 = 16$
(D) $x = y^2$

12. Use the graph of f below to find $f(2)$.

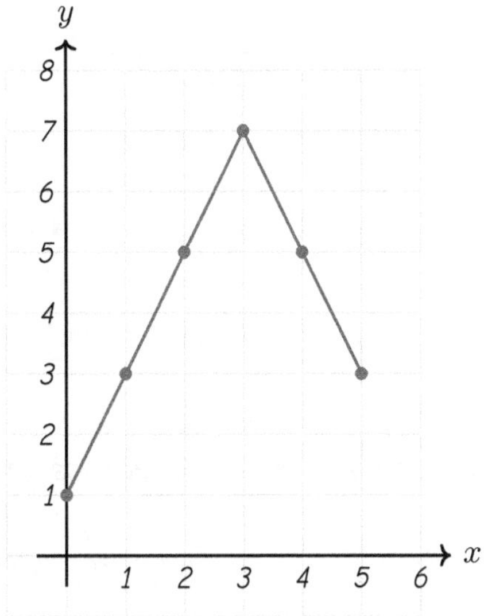

(A) 2 (B) 3

(C) 5 (D) 7

13. Function A: $y = -x + 10$. Function B: $y = -4x + 10$. Which function decreases faster? Write A or B.

Your Answer:

14. Which table shows a linear function?

(A) $(1,1), (2,4), (3,9), (4,16)$ (B) $(0,5), (1,8), (2,11), (3,14)$

(C) $(1,2), (2,4), (3,8), (4,16)$ (D) $(0,1), (1,3), (2,9), (3,27)$

15. *The table below shows a linear function. What is the equation?*

x	0	1	2	3
y	-4	-1	2	5

(A) $y = 3x - 4$

(B) $y = -4x + 3$

(C) $y = -3x + 4$

(D) $y = 3x + 4$

16. *Which graph could represent a person standing still for 5 seconds, then walking at a steady pace?*

(A) A line sloping upward for the entire time

(B) A horizontal line, then a line sloping upward

(C) A line sloping downward, then a horizontal line

(D) A curve going upward the entire time

17. *A triangle is translated 6 units left and 3 units up. A vertex was at $(2, -1)$. What are its new coordinates?*

Your Answer:

18. *True or false: If two figures have the same area, they must be congruent.*

Your Answer:

19. *Point $(3, 4)$ is dilated by a factor of 2 from the origin. What is the image?*

(A) $(5, 6)$

(B) $(6, 8)$

(C) $(1.5, 2)$

(D) $(6, 4)$

20. Two similar pentagons have a scale factor of $\frac{2}{3}$. If the perimeter of the larger pentagon is 45 cm, what is the perimeter of the smaller?

Your Answer:

21. Corresponding angles formed by a transversal and two parallel lines are located:

(A) On opposite sides of the transversal, between the parallel lines

(B) On the same side of the transversal, in matching positions

(C) On opposite sides of the transversal, outside the parallel lines

(D) Adjacent to each other at one intersection

22. Is a triangle with sides 11, 60, 61 a right triangle?

Your Answer:

23. What is the distance between $(-5, 0)$ and $(7, 0)$?

(A) -12

(B) 2

(C) 12

(D) $\sqrt{74}$

24. What is the volume of a cylinder with radius 7 m and height 3 m? Leave your answer in terms of π.

(A) $21\pi\ m^3$

(B) $63\pi\ m^3$

(C) $147\pi\ m^3$

(D) $49\pi\ m^3$

25. A rectangular prism is $10 \times 4 \times 3$ ft. How much wrapping paper is needed to cover it (no overlaps)?

(A) $120\ ft^2$

(B) $164\ ft^2$

(C) $82\ ft^2$

(D) $240\ ft^2$

26. Data: $(1,2),(2,4),(3,5),(4,8),(5,9)$. The dots appear to go upward from left to right. The association is:

(A) negative

(B) positive

(C) no association

(D) nonlinear

27. Data: $(1,3),(2,5),(3,4),(4,6),(5,8)$. A trend line through $(1,3)$ and $(5,8)$ has slope:

(A) $\frac{5}{4}$

(B) $\frac{4}{5}$

(C) 1

(D) 2

28. What is extrapolation?

(A) Making predictions within the range of the data

(B) Making predictions far beyond the range of the data

(C) Removing outliers from the data

(D) Finding the exact equation of the line

29. A survey asks students about their favorite sport (soccer or basketball) and grade (7th or 8th). 20 7th-graders chose soccer. This 20 is a:

(A) marginal frequency

(B) joint frequency

(C) relative frequency

(D) total frequency

30. What is the probability of an impossible event?

(A) 1

(B) 0.5

(C) 0

(D) -1

End of Practice Test 6

Great job finishing the test!

My Score

I got __________ out of 30 questions right.

*Check your answers in the **Answer Key** at the back of the book.*

Review any questions you missed. That's how we learn!

Check Your Score Online!

*Visit **ViewMath Academy** to enter your answers and see which topics you need to review. You can also explore lessons, take quizzes, track your scores, and save your progress!*

viewmath.com/score/8.1.OK.21

*Or go to **viewmath.com/score** and enter code: **8.1.OK.21***

PRACTICE TEST

7

Practice Test 7

30 Questions

Before You Start

- **Read each question carefully** before choosing your answer.
- **Show your work** on scratch paper when you need to.
- **Skip hard questions** and come back to them later.
- **Check your answers** when you're done.
- **Take your time** — there's no rush!

You've Got This!

Do your best and show what you know!

1. *Give an example of an irrational number between* 1 *and* 2.

Your Answer:

2. *A student converts* $0.\overline{36}$ *and gets* $\frac{36}{100}$. *What mistake did the student make?*

(A) *The student divided by* 100 *instead of* 99.

(B) *The student forgot to simplify.*

(C) *The student treated a repeating decimal as a terminating decimal.*

(D) *The student multiplied by* 10 *instead of* 100.

3. *Between which two consecutive integers does* $\sqrt{83}$ *lie?*

Your Answer:

4. *A student claims that* $2\sqrt{5} = \sqrt{10}$. *Is this correct?*

(A) *Yes, because* $2 \times 5 = 10$.

(B) *No,* $2\sqrt{5} = \sqrt{20}$, *not* $\sqrt{10}$.

(C) *Yes, multiplication distributes into the square root.*

(D) *No,* $2\sqrt{5} = 4\sqrt{5}$.

5. *A store buys a backpack for* $\$24$ *and marks it up by* 75%. *What is the selling price?*

Your Answer:

6. *What is the simplified form of* $3^4 \cdot 3^2$?

(A) 3^6

(B) 3^8

(C) 9^6

(D) 3^2

7. Estimate the product $498{,}000{,}000 \times 6{,}200$ by first writing each factor in scientific notation and then computing.

Your Answer:

8. What is $\frac{6.4\times10^9}{3.2\times10^6}$?

(A) 2×10^3
(B) 2×10^{15}
(C) 3.2×10^3
(D) 2×10^{-3}

9. What is the slope of the line through $(1,3)$ and $(4,12)$?

(A) 2
(B) 3
(C) 4
(D) 9

10. Which equation has the steepest line?

(A) $y = x + 1$
(B) $y = 3x - 2$
(C) $y = -5x + 4$
(D) $y = \frac{1}{2}x + 6$

11. How many times can a vertical line cross the graph of a function at most?

Your Answer:

12. A function is defined by the table below.

x	0	1	2	3
$f(x)$	-1	3	7	11

What is $f(0) + f(2)$?

(A) 2
(B) 6
(C) 8
(D) 10

13. Two functions have the same rate of change but different initial values. What is true about their graphs?

(A) They are the same line.
(B) They are parallel lines.
(C) They intersect at the origin.
(D) One is curved and the other is straight.

14. A table shows $(1, 3)$, $(2, 6)$, $(3, 11)$, $(4, 18)$. Is the function linear or nonlinear?

(A) Linear, because y increases as x increases.
(B) Linear, because the first difference is 3.
(C) Nonlinear, because the differences in y are $3, 5, 7$ — not constant.
(D) Nonlinear, because all outputs are positive.

15. Write the equation of a line with slope 6 and y-intercept -3.

Your Answer:

16. A flat (horizontal) section of a graph of height vs. time means:

(A) The height is increasing.
(B) The height is decreasing.
(C) The height is staying the same.
(D) Time has stopped.

17. *A figure is moved* 4 *units to the right and* 2 *units down without changing its size or shape. What transformation is this?*

(A) *Rotation*

(B) *Reflection*

(C) *Translation*

(D) *Dilation*

18. $\triangle ABC \cong \triangle DEF$. *If* $\angle A = 55^\circ$ *and* $\angle B = 80^\circ$, *what is* $\angle F$?

(A) 55°

(B) 80°

(C) 45°

(D) 135°

19. *A triangle has a vertex at* $(8, -4)$. *After a dilation by factor* $\frac{1}{2}$ *from the origin, what is the image of this vertex?*

(A) $(16, -8)$

(B) $(4, -2)$

(C) $(4, -4)$

(D) $(8, -2)$

20. *Triangle* A *has sides* 6, 8, 10. *Triangle* B *has sides* 9, 12, 15. *What is the scale factor from* A *to* B?

Your Answer

21. *In an isosceles triangle, the two base angles are each* 54°. *What is the vertex angle?*

(A) 54°

(B) 72°

(C) 108°

(D) 126°

22. *Is a triangle with sides* 6, 8, 11 *a right triangle?*

(A) *Yes*

(B) *No, because* $6^2 + 8^2 \neq 11^2$

(C) *No, because* $6 + 8 \neq 11$

(D) *Yes, because all sides are different lengths.*

23. *Find the distance between* $(3, 1)$ *and* $(7, 4)$.

Your Answer:

24. *What is the volume of the cone shown below in terms of* π?

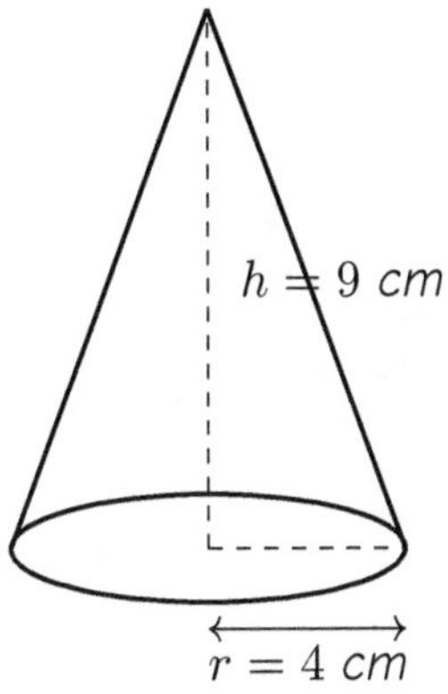

(A) $144\pi\ cm^3$

(B) $48\pi\ cm^3$

(C) $36\pi\ cm^3$

(D) $108\pi\ cm^3$

25. *A rectangular prism is* $10 \times 8 \times h$ *and has surface area* $376\ cm^2$. *Find* h.

Your Answer:

26. *A scatter plot has a strong upward trend. Which* r*-value (correlation) is most likely?*

(A) $r = -0.9$

(B) $r = 0.1$

(C) $r = 0.9$

(D) $r = 0$

27. *Data:* $(1, 8), (2, 7), (3, 5), (4, 4), (5, 2)$. *Draw a trend line and write its equation.*

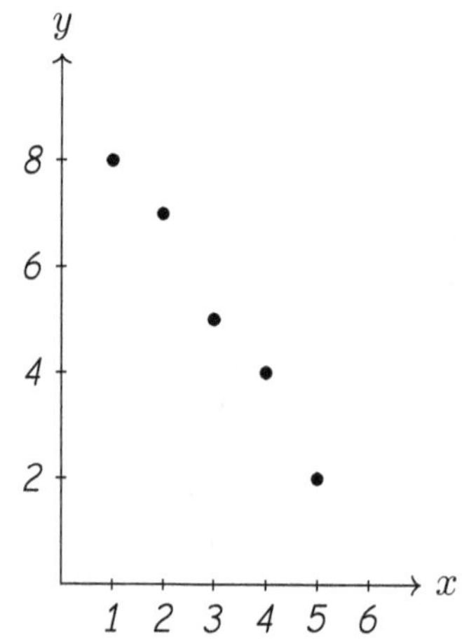

Your Answer:

28. *A linear model is* $y = 2x + 5$. *What does the slope represent?*

(A) The starting value

(B) The rate of change per unit increase in x

(C) The x-intercept

(D) The total value of y

29. *Using the table above, what proportion of pizza lovers are boys?*

(A) $\frac{18}{42}$

(B) $\frac{24}{42}$

(C) $\frac{24}{80}$

(D) $\frac{24}{40}$

30. *A card is drawn from a standard* 52*-card deck. What is* $P(\text{heart})$*?*

(A) $\frac{1}{52}$

(B) $\frac{1}{13}$

(C) $\frac{1}{4}$

(D) $\frac{1}{2}$

★ End of Practice Test 7 ★

Great job finishing the test!

My Score

I got __________ out of 30 questions right.

*Check your answers in the **Answer Key** at the back of the book.*

Review any questions you missed. That's how we learn!

Check Your Score Online!

*Visit **ViewMath Academy** to enter your answers and see which topics you need to review. You can also explore lessons, take quizzes, track your scores, and save your progress!*

viewmath.com/score/8.1.OK.22

*Or go to **viewmath.com/score** and enter code: **8.1.OK.22***

PRACTICE TEST

8

Practice Test 8

30 Questions

Before You Start

- **Read each question carefully** before choosing your answer.
- **Show your work** on scratch paper when you need to.
- **Skip hard questions** and come back to them later.
- **Check your answers** when you're done.
- **Take your time** — there's no rush!

You've Got This!

Do your best and show what you know!

1. Which of the following is a true statement?

(A) Every square root is irrational.
(B) Every integer is irrational.
(C) Every integer is rational.
(D) Every decimal is irrational.

2. The table below shows repeating decimals and their fraction equivalents. Find the missing fraction for $0.\overline{81}$.

Decimal	Unsimplified Fraction	Simplest Form
$0.\overline{27}$	$\frac{27}{99}$	$\frac{3}{11}$
$0.\overline{54}$	$\frac{54}{99}$	$\frac{6}{11}$
$0.\overline{81}$	?	?

Your Answer:

3. Approximate $\sqrt{45}$ to one decimal place.

Your Answer:

4. Which expression equals exactly 6?

(A) $\sqrt{3} \times \sqrt{12}$
(B) $\sqrt{3} + \sqrt{12}$
(C) $\sqrt{6} \times \sqrt{6}$
(D) Both A and C

5. How long will it take for $400 to earn $60 in simple interest at 5% per year?

Your Answer:

6. *Study the number line below. Which expression corresponds to the point marked with a star (★)?*

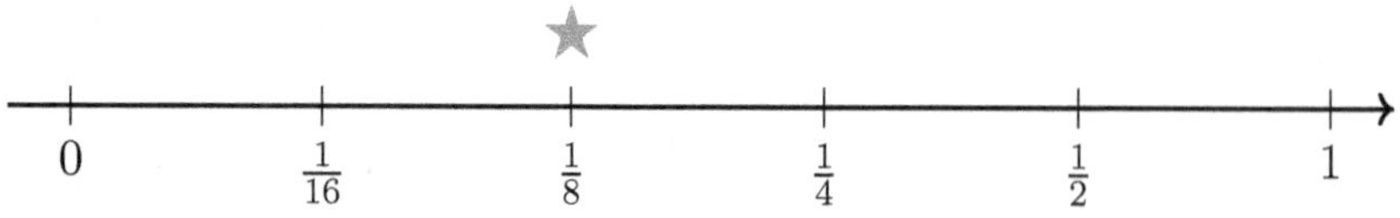

(A) 2^{-2}

(B) 2^{-3}

(C) 2^{-4}

(D) 2^{-1}

7. *A human hair is about 8×10^{-5} meters thick. A spider silk strand is about 4×10^{-6} meters thick. How many times thicker is a human hair than spider silk?*

(A) 2 *times*

(B) 4 *times*

(C) 20 *times*

(D) 200 *times*

8. *A virus has a mass of 9.5×10^{-18} grams. If a sample contains 2×10^{6} viruses, what is the total mass?*

(A) 1.9×10^{-11} *g*

(B) 1.9×10^{-12} *g*

(C) 11.5×10^{-12} *g*

(D) 1.9×10^{-24} *g*

9. *The temperature drops from 68°F to 50°F over 6 hours. What is the rate of change in temperature?*

(A) 3°F *per hour*

(B) −3°F *per hour*

(C) 6°F *per hour*

(D) −6°F *per hour*

10. *A line has equation $y = 4x - 10$. At what value of x does the line cross the x-axis?*

Your Answer:

11. Which graph below represents a function?

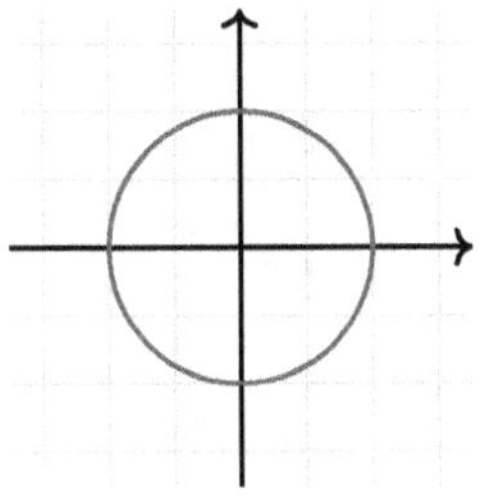

Graph A

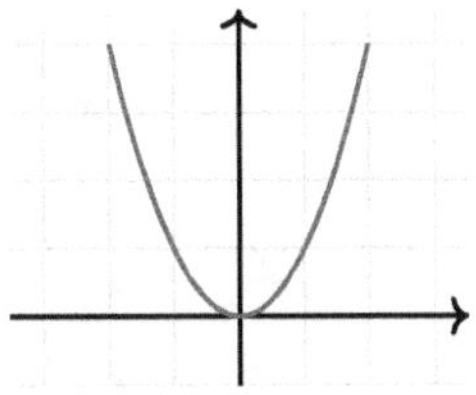

Graph B

(A) Graph A only

(B) Graph B only

(C) Both Graph A and Graph B

(D) Neither Graph A nor Graph B

12. If $f(x) = -x + 8$, what is $f(-3)$?

(A) 5

(B) -5

(C) 11

(D) -11

13. Company A charges $\$20$ plus $\$8$ per hour. Company B charges $\$50$ plus $\$3$ per hour. After how many hours do they charge the same amount?

Your Answer

14. *Look at the graph below. Is the function linear or nonlinear? Explain your reasoning.*

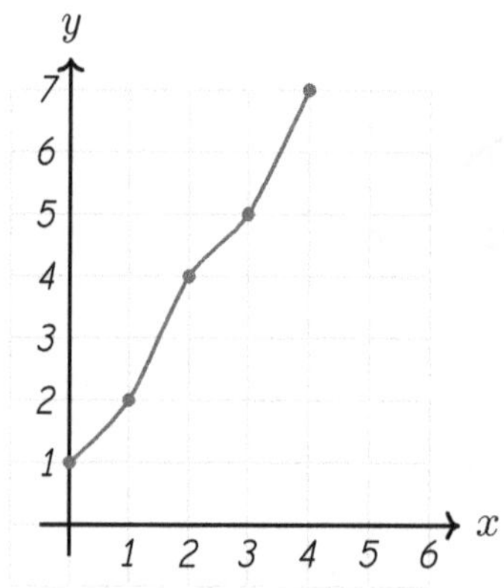

Your Answer:

15. *Find the equation of the line through* $(1, 3)$ *and* $(4, 12)$.

Your Answer:

16. *A graph of savings over time goes up steeply for months 1–3, then less steeply for months 4–6. What changed?*

Your Answer:

17. *Point* $N(0, 4)$ *is rotated* $180°$ *around the origin. What are the coordinates of* N'?

Your Answer:

18. Triangle RST has vertices $R(0,0)$, $S(4,0)$, and $T(0,3)$. Triangle $R'S'T'$ has vertices $R'(0,0)$, $S'(0,4)$, and $T'(-3,0)$. What transformation maps RST to $R'S'T'$?

(A) Translation

(B) Reflection over the x-axis

(C) $90°$ counterclockwise rotation

(D) $90°$ clockwise rotation

19. Dilate the point $(-4, 10)$ by a factor of $\frac{1}{2}$ from the origin. What is the image?

Your Answer:

20. Square $ABCD$ is dilated from the origin by $k = 2$ to form $A'B'C'D'$. Which graph shows the correct image?

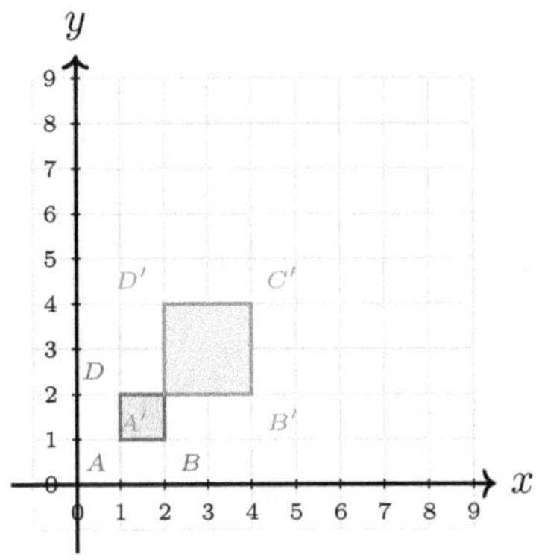

(A) $A'(2,2)$, $B'(4,2)$, $C'(4,4)$, $D'(2,4)$ — as shown

(B) $A'(3,3)$, $B'(4,3)$, $C'(4,4)$, $D'(3,4)$

(C) $A'(1,1)$, $B'(3,1)$, $C'(3,3)$, $D'(1,3)$

(D) $A'(2,1)$, $B'(4,1)$, $C'(4,2)$, $D'(2,2)$

21. A co-interior angle measures $97°$. What is the other co-interior angle?

Your Answer:

22. Which equation represents the Pythagorean Theorem?

(A) $a + b = c$
(B) $a^2 + b^2 = c^2$
(C) $a^2 \times b^2 = c^2$
(D) $2a + 2b = c$

23. What is the distance between $(1, 3)$ and $(4, 7)$?

(A) 7
(B) 25
(C) $\sqrt{7}$
(D) 5

24. Find the volume of a sphere with radius 6 in. Leave your answer in terms of π.

Your Answer:

25. A cylinder has a lateral surface area of 88π in^2 and a radius of 4 in. What is the height?

Your Answer:

26. In a scatter plot, several data points bunch together near $(3, 20)$. This grouping is called:

(A) an outlier
(B) a gap
(C) a cluster
(D) a negative association

27. Data: $(1, 2), (2, 4), (3, 7), (4, 8), (5, 11)$. Estimate the slope of a reasonable trend line.

Your Answer:

28. In the model $y = 6x + 12$, if x increases by 3 units, how much does y increase?

(A) 3
(B) 6
(C) 18
(D) 30

29. Using the table from q19, what percentage of boys prefer reading? What percentage of girls prefer reading? Is there an association?

Your Answer:

30. $P(\text{event}) = \frac{3}{4}$. What is $P(\text{not event})$?

(A) $\frac{3}{4}$
(B) $\frac{1}{4}$
(C) $\frac{1}{2}$
(D) 0

End of Practice Test 8

Great job finishing the test!

My Score

I got __________ out of 30 questions right.

Check your answers in the **Answer Key** at the back of the book.

Review any questions you missed. That's how we learn!

Check Your Score Online!

Visit **ViewMath Academy** to enter your answers and see which topics you need to review. You can also explore lessons, take quizzes, track your scores, and save your progress!

viewmath.com/score/8.1.OK.23

Or go to **viewmath.com/score** and enter code: **8.1.OK.23**

PRACTICE TEST

9

Practice Test 9

30 Questions

Before You Start

- **Read each question carefully** before choosing your answer.
- **Show your work** on scratch paper when you need to.
- **Skip hard questions** and come back to them later.
- **Check your answers** when you're done.
- **Take your time** — there's no rush!

You've Got This!

Do your best and show what you know!

1. Which of the following could NOT be the decimal expansion of a rational number?

(A) 2.500

(B) $0.\overline{9}$

(C) $1.41421356\ldots$ (non-repeating)

(D) $0.\overline{36}$

2. What is $0.\overline{4}$ written as a fraction?

(A) $\frac{4}{10}$

(B) $\frac{4}{9}$

(C) $\frac{2}{5}$

(D) $\frac{1}{4}$

3. A student says $\sqrt{45}$ is between 6 and 7. Is the student correct?

(A) Yes, because $6^2 = 36$ and $7^2 = 49$.

(B) No, it is between 5 and 6.

(C) No, it is between 7 and 8.

(D) No, $\sqrt{45} = 9$ exactly.

4. Which is larger: $\sqrt{2} + \sqrt{3}$ or $\sqrt{10}$?

(A) $\sqrt{2} + \sqrt{3}$, because $2 + 3 = 5 > \sqrt{10}$

(B) $\sqrt{10}$, because $10 > 5$

(C) $\sqrt{10}$, because $\sqrt{10} \approx 3.16 > 3.15 \approx \sqrt{2} + \sqrt{3}$

(D) They are equal.

5. You deposit $\$500$ at 4% per year with compound interest (compounded yearly). How much do you have after 2 years?

(A) $\$540.00$

(B) $\$540.80$

(C) $\$580.00$

(D) $\$520.00$

6. *Simplify* $7^5 \cdot 7^{-5}$.

(A) 7^{25}

(B) 7^{10}

(C) 0

(D) 1

7. *The mass of a proton is approximately* 1.67×10^{-27} *kilograms. What does the exponent* -27 *tell you?*

(A) *The proton weighs* 27 *kilograms*

(B) *The decimal point moves* 27 *places to the right*

(C) *The number is extremely small, with* 26 *zeros after the decimal point before the* 167

(D) *The proton weighs* -27 *kilograms*

8. *The diagram below shows the input and output of a machine that divides. Find the output value. Write your answer in scientific notation.*

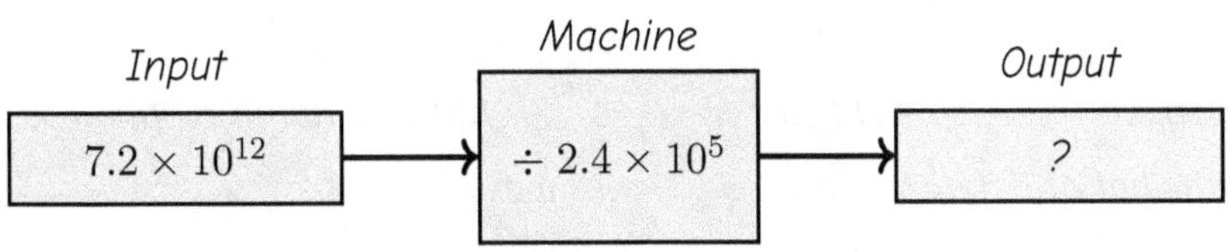

Your Answer:

9. *Find the slope through* $(0, -2)$ *and* $(3, 7)$.

(A) 3

(B) -3

(C) $\frac{5}{3}$

(D) $\frac{9}{3}$

10. *Read the graph and write the equation of the line in slope-intercept form.*

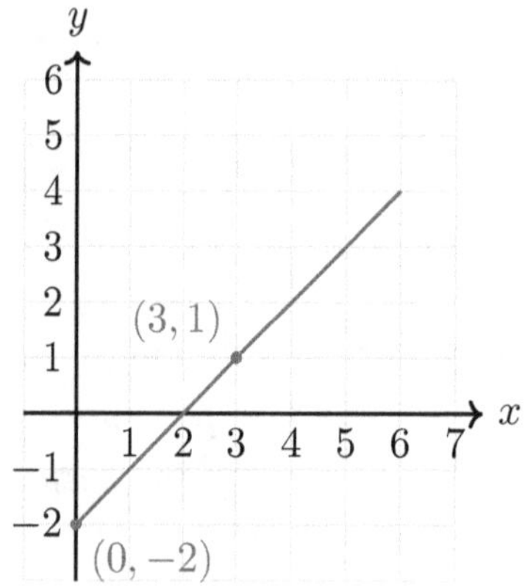

Your Answer:

11. *Which of these real-world examples is a function?*

(A) *A person's age mapped to their height (a person can be the same height at different ages)*

(B) *A student's name mapped to their student ID number*

(C) *A city mapped to all the people who live there*

(D) *A month mapped to every holiday in that month*

12. *If* $f(x) = 3x + 1$, *what is* $f(4)$*?*

(A) 7

(B) 12

(C) 13

(D) 15

13. *Two hikers start a trail. Hiker A walks* 3 *miles per hour and starts* 2 *miles from the trailhead. Hiker B is described by* $y = 4x$. *Who walks faster?*

(A) *Hiker A*

(B) *Hiker B*

(C) *They walk at the same speed.*

(D) *Cannot be determined.*

14. Which is a characteristic of a linear function?

(A) Its graph is a curve.

(B) Its rate of change is constant.

(C) Its rate of change varies.

(D) It always passes through the origin.

15. A line passes through $(-1, 2)$ and $(3, 10)$. What is the slope?

(A) 2

(B) 3

(C) 4

(D) 8

16. A graph shows temperature over a day. From 6 AM to noon, the graph goes up steeply. From noon to 3 PM, it goes up slowly. What can you conclude?

(A) The temperature decreased between noon and 3 PM.

(B) The temperature increased faster in the morning than in the afternoon.

(C) The temperature was constant from noon to 3 PM.

(D) The temperature increased at the same rate all day.

17. Point $P(2, 0)$ is rotated $90°$ counterclockwise around the origin. Where does P' land?

(A) $(0, -2)$

(B) $(0, 2)$

(C) $(-2, 0)$

(D) $(2, 0)$

18. Which pair of figures is NOT necessarily congruent?

(A) A triangle and its reflection

(B) A rectangle and its translation

(C) Two rectangles with the same perimeter

(D) A square and its $90°$ rotation

19. *Point $A(0,-5)$ is rotated 90° counterclockwise around the origin. Where does A' land?*

(A) $(5,0)$
(B) $(0,5)$
(C) $(-5,0)$
(D) $(0,-5)$

20. *A triangle has an area of 20 cm^2. It is dilated by $k=3$. What is the area of the image?*

(A) 60 cm^2
(B) 180 cm^2
(C) 120 cm^2
(D) 6.67 cm^2

21. *Parallel lines ℓ and m are cut by transversal t. If a pair of alternate interior angles are $(4x)^\circ$ and 80°, what is x?*

(A) 320
(B) 20
(C) 25
(D) 10

22. *In a right triangle, the legs are 6 and 8. What is the hypotenuse?*

(A) 14
(B) 10
(C) 48
(D) $\sqrt{48}$

23. *What is the distance between $(0,0)$ and $(1,1)$?*

(A) 2
(B) 1
(C) $\sqrt{2}$
(D) 0

24. *What is the volume formula for a cone?*

(A) $V = \pi r^2 h$

(B) $V = \frac{1}{2}\pi r^2 h$

(C) $V = \frac{1}{3}\pi r^2 h$

(D) $V = \frac{4}{3}\pi r^3$

25. *What is the surface area of the rectangular prism shown?*

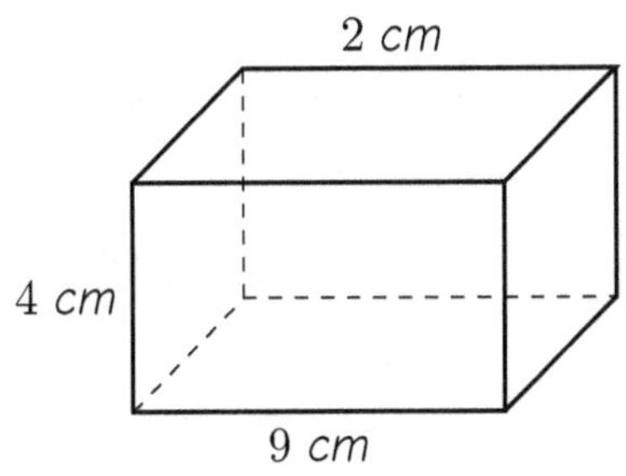

(A) 72 cm^2

(B) 124 cm^2

(C) 148 cm^2

(D) 100 cm^2

26. *True or false: A scatter plot can show both clustering and outliers at the same time.*

(A) *True — they describe different features of the data.*

(B) *False — data has either clusters or outliers, not both.*

(C) *True — but only if there is no association.*

(D) *False — outliers are always in clusters.*

27. A trend line is drawn through the scatter plot. What is the approximate slope of the line?

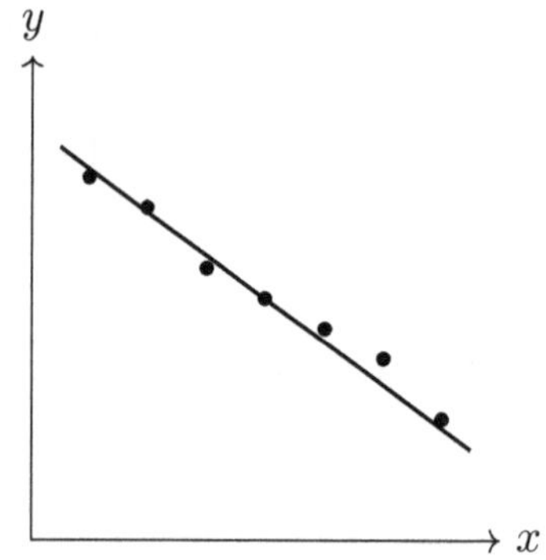

(A) $\approx -\frac{5}{7}$

(B) $\approx \frac{5}{7}$

(C) ≈ -2

(D) ≈ 1

28. A model gives $y = 5x + 20$ where x is weeks and y is savings (dollars). If $x = 8$, what is the predicted savings?

(A) $\$45$

(B) $\$60$

(C) $\$68$

(D) $\$40$

29. Which of these is NOT a question you can answer with a two-way table?

(A) What fraction of boys prefer soccer?

(B) Is there an association between grade and preferred lunch?

(C) What is the mean score on the math test?

(D) How many 8th graders chose art?

30. Explain the difference between independent and dependent events with an example.

Your Answer:

★ End of Practice Test 9 ★

Great job finishing the test!

My Score

I got __________ out of 30 questions right.

Check your answers in the ***Answer Key*** *at the back of the book.*

Review any questions you missed. That's how we learn!

Check Your Score Online!

Visit ***ViewMath Academy*** *to enter your answers and see which topics you need to review. You can also explore lessons, take quizzes, track your scores, and save your progress!*

viewmath.com/score/8.1.OK.24

Or go to ***viewmath.com/score*** *and enter code:* ***8.1.OK.24***

PRACTICE TEST

10

Practice Test 10

30 Questions

Before You Start

- **Read each question carefully** before choosing your answer.
- **Show your work** on scratch paper when you need to.
- **Skip hard questions** and come back to them later.
- **Check your answers** when you're done.
- **Take your time** — there's no rush!

You've Got This!

Do your best and show what you know!

1. Look at the Venn diagram below. In which region does $\sqrt{5}$ belong?

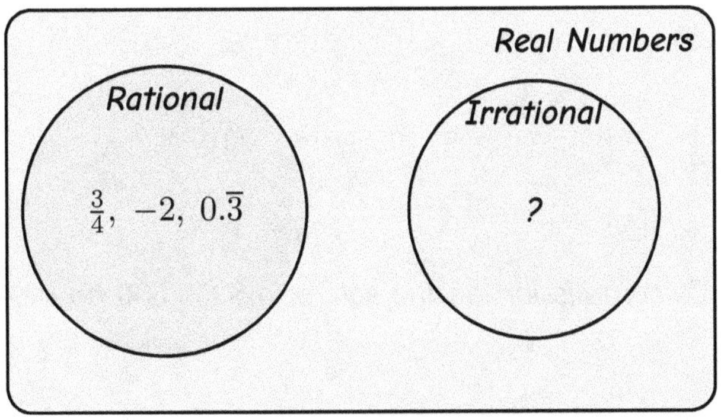

(A) Rational region, because $\sqrt{5} = 2.5$

(B) Irrational region, because 5 is not a perfect square

(C) Rational region, because $\sqrt{5}$ is a square root

(D) Neither region, because $\sqrt{5}$ is not a real number

2. The steps below show a student's work for converting $0.\overline{2}$ to a fraction. Which step contains the error?

Step	Work
1	Let $x = 0.222\ldots$
2	$10x = 2.222\ldots$
3	$10x - x = 2.222\ldots - 0.222\ldots$
4	$9x = 2$
5	$x = \frac{2}{10}$

(A) Step 2

(B) Step 3

(C) Step 4

(D) Step 5

3. Between which two consecutive integers does $\sqrt{50}$ lie?

(A) 6 and 7

(B) 7 and 8

(C) 8 and 9

(D) 24 and 26

4. *Order from least to greatest:* π, $\sqrt{10}$, 3.2.

(A) $\pi,\ 3.2,\ \sqrt{10}$

(B) $3.2,\ \pi,\ \sqrt{10}$

(C) $\sqrt{10},\ \pi,\ 3.2$

(D) $\pi,\ \sqrt{10},\ 3.2$

5. *A real estate agent earns a 3% commission on the sale of a \$250,000 house. How much commission does the agent earn?*

(A) \$750

(B) \$7,500

(C) \$75,000

(D) \$25,000

6. *Look at the partially completed exponent chart below. Find the values of A and B, then compute $A \times B$.*

Power	3^3	3^2	3^1	3^0
Value	A	9	B	1

Your Answer:

7. *Write* 0.0000059 *in scientific notation.*

Your Answer:

8. *A country has 3×10^8 people and an area of 6×10^6 square kilometers. What is the population density (people per square kilometer)?*

Your Answer:

9. *A line goes down from left to right. Its slope must be:*

(A) *Positive*

(B) *Negative*

(C) *Zero*

(D) *Undefined*

10. *A line passes through the origin and* $(3, 12)$. *What is its equation?*

(A) $y = 3x$

(B) $y = 12x$

(C) $y = 4x$

(D) $y = \frac{1}{4}x$

11. *Use the mapping diagram below. List all the inputs that are part of this relation.*

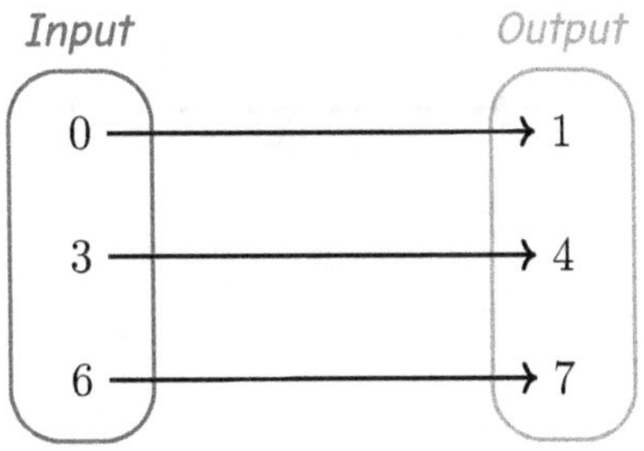

Your Answer:

12. *If* $f(x) = 6x - 4$, *for which value of* x *is* $f(x) = 0$?

(A) $x = 0$

(B) $x = \frac{2}{3}$

(C) $x = 1$

(D) $x = 4$

13. *The graph below shows Function F. Function G is $y = 3x + 2$. Find the rate of change of each function and state which is greater.*

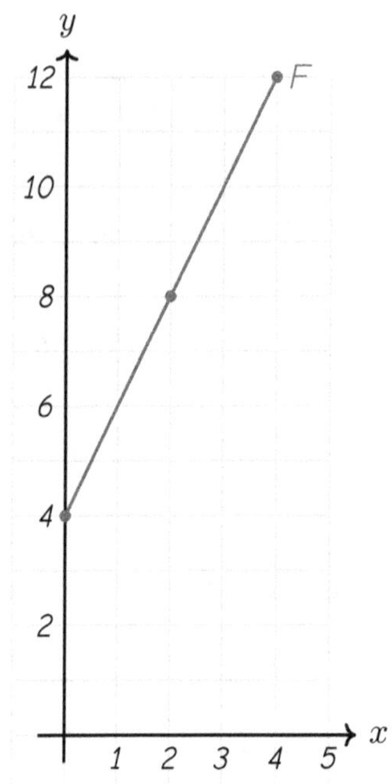

Your Answer

14. *What is the form of a linear function?*

(A) $y = ax^2 + bx + c$

(B) $y = mx + b$

(C) $y = a^x$

(D) $y = \frac{a}{x}$

15. *Sam has \$120 and spends \$15 per day. Which function models the money remaining after x days?*

(A) $y = 15x + 120$

(B) $y = -15x + 120$

(C) $y = 120x - 15$

(D) $y = -120x + 15$

16. *A graph of amount of gas in a car over time goes downward. Is the function increasing, decreasing, or constant?*

Your Answer:

17. *A student says that a reflection changes the area of a figure. Is this correct?*

(A) Yes, because the figure flips.

(B) Yes, because one dimension reverses.

(C) No, because reflections preserve all measurements.

(D) No, but reflections change the perimeter.

18. *Triangle ABC is shown on the grid below. $A'B'C'$ is congruent to ABC and formed by reflecting ABC over the x-axis. What are the coordinates of C'?*

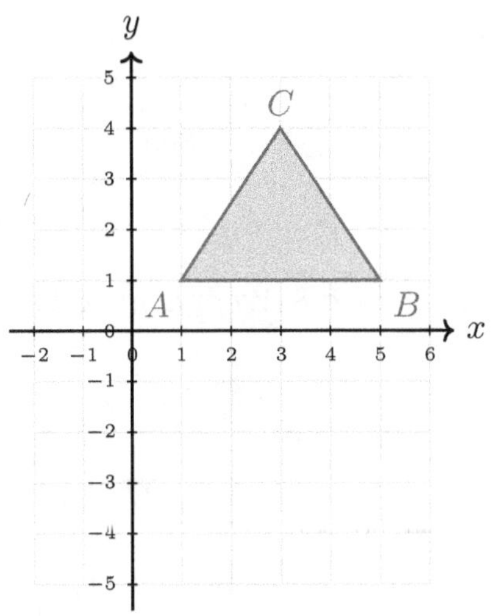

Your Answer:

19. *A point is translated by $(3, -2)$ and then reflected over the y-axis. If the original point is $(1, 4)$, what is the final image?*

(A) $(-4, 2)$

(B) $(4, 2)$

(C) $(-4, -2)$

(D) $(4, -2)$

20. *A figure is dilated by $k = 0.25$. What happens to the figure?*

(A) *It gets 4 times larger.*

(B) *It stays the same size.*

(C) *It shrinks to $\frac{1}{4}$ of its original size.*

(D) *It disappears.*

21. *What is the sum of the interior angles of any triangle?*

(A) $90°$

(B) $180°$

(C) $270°$

(D) $360°$

22. *A right triangle has hypotenuse 17 and one leg 8. What is the other leg?*

(A) 9

(B) 15

(C) 25

(D) $\sqrt{225}$

23. *A rectangle has corners at $(0, 0)$ and $(9, 12)$. What is the length of the diagonal?*

Your Answer

24. A sphere has the dimensions shown. What is its volume? Use $\pi \approx 3.14$

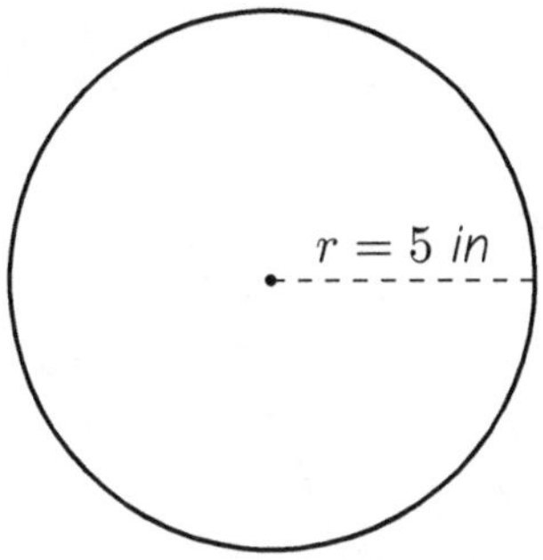

(A) 78.5 in^3
(B) 523.3 in^3
(C) 392.5 in^3
(D) 314.0 in^3

25. A cube has surface area 150 cm^2. What is the edge length?

(A) 25 cm
(B) 5 cm
(C) 6 cm
(D) 10 cm

26. A scatter plot of age (x) versus height (y) for people aged 0–80 would likely show what shape?

Your Answer:

27. When drawing an informal line of best fit, about how many data points should be above the line?

(A) All of them
(B) None of them
(C) About half
(D) Exactly one

28. A scatter plot with trend line $y = 2x + 1$ is shown. What is the predicted value at $x = 4$?

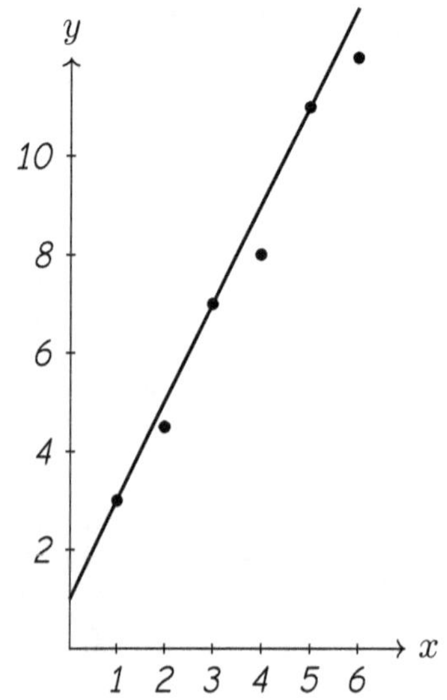

(A) 7

(B) 8

(C) 9

(D) 10

29.

	Pizza	Tacos	Total
Boys	24	16	40
Girls	18	22	40
Total	42	38	80

What is the relative frequency of boys who prefer pizza out of all 80 students?

(A) 0.30

(B) 0.60

(C) 0.20

(D) 0.525

30. *A bag has 5 red and 3 blue marbles. One marble is drawn and replaced. Then another is drawn. What is P(red, then blue)?*

(A) $\frac{15}{64}$

(B) $\frac{15}{56}$

(C) $\frac{5}{8}$

(D) $\frac{3}{8}$

★ End of Practice Test 10 ★

Great job finishing the test!

My Score

I got __________ out of 30 questions right.

*Check your answers in the **Answer Key** at the back of the book.*

Review any questions you missed. That's how we learn!

Check Your Score Online!

*Visit **ViewMath Academy** to enter your answers and see which topics you need to review. You can also explore lessons, take quizzes, track your scores, and save your progress!*

viewmath.com/score/8.1.OK.25

*Or go to **viewmath.com/score** and enter code: **8.1.OK.25***

Answer Key & Explanations

Answer Key

First try each test on your own, then check your work here.

Practice Test 1 — Answer Key

1. False
2. C
3. D
4. B
5. B
6. -1
7. D
8. B
9. 3
10. $y = 2x - 3$
11. C
12. 3
13. Yes
14. C
15. 4
16. A
17. Translation, reflection, rotation
18. $55°$
19. B
20. B
21. C
22. B
23. A
24. C
25. B
26. B
27. B
28. At $x = 5$: $y = 30$. At $x = 30$: $y = -45$. The prediction at $x = 5$ is more reliable because it is within the data ran
29. C
30. $\frac{4}{52} \times \frac{3}{51} = \frac{12}{2652} = \frac{1}{221}$

Time to Learn!

Review the explanations below, ***especially for the questions you missed.***

Understanding why each answer is correct builds stronger problem-solving skills.

Tip: *Circle any questions you got wrong, then read their explanation carefully.*

Practice Test 1 — Detailed Explanations

1 $\frac{22}{7} \approx 3.142857\ldots$ *is a rational approximation of* π*, but* $\pi = 3.14159265\ldots$ *is irrational. They are close but not equal.*

2 $0.\overline{123}$ *has a 3-digit repeating block, so you should multiply by* $10^3 = 1000$*, not* 100*.*

3 $3.74^2 = 13.9876$ *(under) and* $3.75^2 = 14.0625$ *(just over).* 14.0625 *is very close to* 14*, so* 3.75 *is the best next estimate.*

4 $\sqrt{7} \approx 2.646$*, so* $5 - \sqrt{7} \approx 5 - 2.646 = 2.354 \approx 2.4$*.*

5 *Tax* $= 65 \times 0.08 = 5.20$*. Total* $= 65 + 5.20 = \$70.20$*.*

6 (-1) *raised to an odd power is* -1*. Since* 15 *is odd,* $(-1)^{15} = -1$*.*

7 *The Sun has an exponent of* 6 *(the largest), so* $1.4 \times 10^6 = 1{,}400{,}000$ *km is the greatest diameter.*

8 $\frac{8\times10^4}{2\times10^2} = 4 \times 10^2 = 400$ *minutes.*

9 $m = \frac{11-(-1)}{7-3} = \frac{12}{4} = 3$*.*

10 $m = \frac{9-1}{6-2} = 2$*. Use* $(2,1)$*:* $1 = 2(2) + b$*, so* $b = -3$*. Equation:* $y = 2x - 3$*.*

11 *At* $x = 3$ *there are two different* y*-values (1 and 5). A vertical line at* $x = 3$ *would cross two points, so this is not a function.*

12 *Follow the dashed line at* $y = 6$ *until it meets the graph. That point is* $(3,6)$*, so* $x = 3$*.*

13 *Lines with the same slope but different* y*-intercepts are parallel.*

14 A linear function has the form $y = mx + b$. Only $y = 4x - 7$ fits this form with $m = 4$ and $b = -7$.

15 $\frac{25-9}{6-2} = \frac{16}{4} = 4$.

16 Walking increases distance. Waiting means distance stays the same (constant). Walking again increases distance further.

17 The three rigid transformations are translation (slide), reflection (flip), and rotation (turn). They all preserve size and shape.

18 $\angle R = 180 - 72 - 53 = 55°$. Since $\angle R$ corresponds to $\angle U$, $\angle U = 55°$.

19 Original sides: 6, 8, and $\sqrt{6^2 + 8^2} = 10$. Perimeter $= 6 + 8 + 10 = 24$. Dilation by $\frac{1}{2}$ multiplies every length by $\frac{1}{2}$, so the new perimeter is $24 \times \frac{1}{2} = 12$.

20 Divide the image coordinates by the original: $\frac{12}{4} = 3$ and $\frac{-6}{-2} = 3$. The scale factor is 3.

21 Corresponding angles (same position at each intersection) are equal when the lines are parallel.

22 $c^2 = 5^2 + 5^2 = 25 + 25 = 50$, so $c = \sqrt{50}$.

23 $d = \sqrt{(4 - (-2))^2 + (9 - 1)^2} = \sqrt{6^2 + 8^2} = \sqrt{36 + 64} = \sqrt{100} = 10$.

24 The sphere formula $V = \frac{4}{3}\pi r^3$ uses only the radius. There is no separate height for a sphere.

25 $200\pi = 2\pi(25) + 2\pi(5)h = 50\pi + 10\pi h$. So $150\pi = 10\pi h$ and $h = 15$ cm.

26 The dots go from upper-left to lower-right in a roughly straight pattern. This is a negative linear association.

27 *Outliers should generally be ignored when drawing the line of best fit, as they don't represent the overall trend.*

28 $y = -3(5) + 45 = 30$ *(interpolation).* $y = -3(30) + 45 = -45$ *(extrapolation — unreliable and gives a negative value that may not make sense).*

29 $\frac{25}{40} = 0.625 = 62.5\%$.

30 *First ace:* $\frac{4}{52}$. *Second ace (without replacement):* $\frac{3}{51}$. *Multiply:* $\frac{12}{2652} = \frac{1}{221}$.

Practice Test 2 — Answer Key

1 C

2 Multiply by 100; $100x = 54.\overline{54}$; $99x = 54$; $x = \frac{54}{99} = \frac{6}{11}$

3 A: ≈ 4.2, B: ≈ 6.3, C: ≈ 8.5

4 C

5 B

6 B

7 A

8 4×10^{-6}

9 B

10 $-\frac{5}{2}$

11 C

12 3

13 Function A: rate $= 3$, initial $= 20$. Function B: rate $= 5$, initial $= 5$. At $x = 10$, $B = 55$ is greater than $A = 50$.

14 $y = x^2$

15 B

16 C

17 C

18 B

19 $(1, 2)$

20 C

21 A

22 No

23 B

24 100π cm^3

25 A

26 Positive linear association. Outlier at approximately $(1, 7)$ — far above the trend.

27 C

28 C

29 B

30 C

Time to Learn!

*Review the explanations below, **especially for the questions you missed**.*

Understanding why each answer is correct builds stronger problem-solving skills.

***Tip:** Circle any questions you got wrong, then read their explanation carefully.*

Practice Test 2 — Detailed Explanations

1. $P = \sqrt{2}$ *is irrational (since* 2 *is not a perfect square) and* $Q = 3$ *is an integer, which is rational.*

2. *The repeating block has* 2 *digits, so multiply by* 100*. Then* $100x - x = 99x = 54$*, giving* $x = \frac{54}{99} = \frac{6}{11}$.

3. *Side* $= \sqrt{Area}$*.* $\sqrt{18} \approx 4.24 \approx 4.2$*.* $\sqrt{40} \approx 6.32 \approx 6.3$*.* $\sqrt{72} \approx 8.49 \approx 8.5$*.*

4. $\sqrt{5} \approx 2.236$ *and* $\sqrt{3} \approx 1.732$*. Adding:* $2.236 + 1.732 = 3.968 \approx 4.0$*.*

5. $I = Prt$*, so* $r = \frac{I}{Pt} = \frac{96}{800 \times 4} = \frac{96}{3200} = 0.03 = 3\%$*.*

6. $\frac{8^6}{8^6} = 8^{6-6} = 8^0 = 1$*. The expression equals* 8^0*.*

7. *Move the decimal* 4 *places right:* $0.00072 = 7.2 \times 10^{-4}$*.*

8. $8 \times 5 = 40$ *and* $10^{-4} \times 10^{-3} = 10^{-7}$*. Then* $40 \times 10^{-7} = 4 \times 10^{-6}$*.*

9. *Rate of change* $= \frac{16-4}{6} = \frac{12}{6} = 2$ *cm per week.*

10 In slope-intercept form $y = mx + b$, the slope $m = -\frac{5}{2}$.

11 Find the ordered pair with input 0: $(0, 1)$. The output is 1.

12 $f(2) - f(1) = 7 - 4 = 3$.

13 A: slope 3, $b = 20 \Rightarrow A(10) = 3(10) + 20 = 50$. B: slope 5, $b = 5 \Rightarrow B(10) = 5(10) + 5 = 55$. Function B is greater at $x = 10$.

14 Any equation where x has an exponent other than 1 (such as x^2, x^3, $\frac{1}{x}$, $\sqrt{x}$) is nonlinear.

15 Slope $= \frac{18-6}{4-1} = \frac{12}{3} = 4$.

16 Filling: water level rises (increasing). Soaking: level stays the same (constant). Draining: level drops (decreasing).

17 Reflecting over the y-axis flips the sign of the x-coordinate while keeping y the same: $(7, -2) \to (-7, -2)$.

18 When triangles are congruent, all corresponding sides are equal. Since AB corresponds to DE, $DE = 8$ cm.

19 Add the translation: $(4 + (-3),\ -6 + 8) = (1, 2)$.

20 Check ratios: $\frac{6}{3} = 2$ but $\frac{9}{5} = 1.8$. Since $2 \neq 1.8$, the sides are not proportional and the rectangles are not similar.

21 The exterior angle adjacent to the third angle equals the sum of the two non-adjacent interior angles: $37 + 53 = 90°$.

22 $5^2 + 10^2 = 25 + 100 = 125$, but $12^2 = 144$. Since $125 \neq 144$, it is not a right triangle.

23 The points are on the x-axis, so the distance is just the horizontal difference: $|5-0|=5$.

24 $V=\frac{1}{3}\pi(25)(12)=\frac{300\pi}{3}=100\pi\ cm^3$.

25 $SA=2\pi(9)+2\pi(3)(12)=56.52+226.08=282.6\ cm^2$.

26 Direction: positive (up-right). Shape: linear. The point $(1,7)$ is an outlier (most students who studied 1 hour scored around 3, not 7).

27 A line of best fit (trend line) is a straight line drawn to approximate the overall trend, coming close to most points.

28 A slope of 0 means no change in y as x changes; the line is horizontal.

29 60% of boys chose band vs. 45% of girls. The difference in conditional proportions shows an association.

30 2 coin outcomes $\times$ 6 die outcomes $=12$ total outcomes.

Practice Test 3 — Answer Key

1 C | 2 C | 3 B | 4 C | 5 A | 6 B | 7 D | 8 A | 9 B | 10 B

11 C | 12 18 | 13 B | 14 Nonlinear | 15 B | 16 C | 17 D | 18 C | 19 A

20 C | 21 C | 22 8 | 23 C | 24 A | 25 B

26 Yes. Example: population growth over time may curve upward (exponential), showing a positive but nonlinear a

27 B | 28 B | 29 B | 30 C

Time to Learn!

Review the explanations below, ***especially for the questions you missed.***

Understanding why each answer is correct builds stronger problem-solving skills.

Tip: *Circle any questions you got wrong, then read their explanation carefully.*

Practice Test 3 — Detailed Explanations

1 $0.\overline{81}$ is a repeating decimal, so it can be written as a fraction: $0.\overline{81} = \frac{81}{99} = \frac{9}{11}$. The others are irrational.

2 Let $x = 0.\overline{27}$. Then $100x = 27.\overline{27}$. Subtract: $99x = 27$, so $x = \frac{27}{99} = \frac{3}{11}$.

3 $7.7^2 = 59.29$ and $7.8^2 = 60.84$. So $\sqrt{60} \approx 7.75$, making 8 a less accurate estimate. 7.7 or 7.8 is better.

4 $4\sqrt{2} \approx 4 \times 1.414 = 5.656$ and $\sqrt{30} \approx 5.477$. Since $5.656 > 5.477$, $4\sqrt{2}$ is larger.

5 Discount: $35 \times 0.20 = 7$, sale price $= 28$. Tax: $28 \times 0.08 = 2.24$. Final: $28 + 2.24 = \$30.24$.

6 Using the quotient rule: $\frac{5^7}{5^3} = 5^{7-3} = 5^4$.

7 $2.3 \times 10^8 - 1.5 \times 10^8 = 0.8 \times 10^8 = 8 \times 10^7$ km. Both A and B are correct representations.

8 Same exponent, so add the coefficients: $5.2 + 3.8 = 9.0$. Answer: 9×10^6.

9 $\frac{270-120}{5-2} = \frac{150}{3} = 50$ mph.

10 *The rate (slope) is \$2/mile and the starting cost (y-intercept) is \$3:* $y = 2x + 3$.

11 *In a function, the first coordinate* x *is the input and the second coordinate* y *is the output.*

12 $f(3) = 4(3) + 1 = 13$. $f(1) = 4(1) + 1 = 5$. *Sum:* $13 + 5 = 18$.

13 *Function A earns \$12/week* ($m = 12$). *Function B earns \$15/week* ($m = 15$). *Since* $15 > 12$, *Function B earns more per week.*

14 *The variable* x *is raised to the 3rd power. For linear,* x *must have exponent* 1.

15 *No activation fee means* $b = 0$. *The monthly cost is \$35, so* $m = 35$. *The function is* $y = 35x$.

16 *The four sections are: increasing, decreasing, increasing, constant.*

17 $(3, 2) \to (-3, -2)$ *follows the rule* $(x, y) \to (-x, -y)$, *which is a* $180°$ *rotation around the origin.*

18 *A dilation by a factor other than* 1 *changes the size of the figure, so it does not preserve congruence.*

19 *Apply the rule:* $(-1, 6) \to (6, -(-1)) = (6, 1)$.

20 *All squares have four* $90°$ *angles. Any two squares can be mapped by a dilation since the ratio of their sides is constant. So all squares are similar.*

21 *The exterior angle equals the sum of the two non-adjacent interior angles, so it's greater than either one of them individually.*

22 $x^2 = 17^2 - 15^2 = 289 - 225 = 64$, *so* $x = 8$.

23 The side length is 6. The diagonal of a square with side s is $s\sqrt{2} = 6\sqrt{2} \approx 8.49$.

24 Radius $= 12$. $V = \frac{4}{3}(3.14)(12^3) = \frac{4}{3}(3.14)(1728) = \frac{4}{3}(5425.92) \approx 7234.6\ cm^3$.

25 Base area $= 12^2 = 144$. Lateral SA $= 360 - 144 = 216$. $\frac{1}{2}(48)\ell = 216$, so $24\ell = 216$ and $\ell = 9$ in.

26 An association can be positive (both increase) and nonlinear (the trend is curved, not straight).

27 A straight line of best fit only works for data with a linear pattern. Curved data needs a different model.

28 $y = -4(10) + 100 = -40 + 100 = 60$.

29 The intersection of Girls row and Dog column is 12.

30 $P(6) = \frac{1}{6}$. Independent: $P = \frac{1}{6} \times \frac{1}{6} = \frac{1}{36}$.

Practice Test 4 — Answer Key

1 D | 2 $\frac{53}{90}$ | 3 C | 4 6 | 5 B | 6 C | 7 C | 8 1×10^6 | 9 C

10 B | 11 C | 12 6 | 13 B | 14 Nonlinear, each output is x^3. | 15 A | 16 B | 17 C

18 C

19 B | 20 B | 21 B | 22 B | 23 C | 24 $300\ cm^3$ | 25 A | 26 C | 27 C | 28 B

29 Yes. Of instrument players, $\frac{15}{40} = 37.5\%$ play a sport. Of non-instrument players, $\frac{30}{40} = 75\%$ play a sport. The large differer

30 A

Time to Learn!

*Review the explanations below, **especially for the questions you missed**.*

Understanding why each answer is correct builds stronger problem-solving skills.

***Tip:** Circle any questions you got wrong, then read their explanation carefully.*

Practice Test 4 — Detailed Explanations

1 $\sqrt{7}$ *is irrational because 7 is not a perfect square. The other choices are all rational:* $\frac{5}{8} = 0.625$, $0.\overline{6} = \frac{2}{3}$, *and* $\sqrt{9} = 3$.

2 *Let* $x = 0.5888\ldots$. *Then* $10x = 5.888\ldots$ *and* $100x = 58.888\ldots$. *Subtract:* $90x = 53$, *so* $x = \frac{53}{90}$.

3 $3^2 = 9$ *and* $4^2 = 16$. *Since* $9 < 15 < 16$, *we have* $3 < \sqrt{15} < 4$. *Therefore* $\sqrt{15} > 3$ *is true. In fact,* $\sqrt{15} \approx 3.87$.

4 $\sqrt{2} \times \sqrt{18} = \sqrt{2 \times 18} = \sqrt{36} = 6$.

5 $I = Prt = 600 \times 0.05 \times 3 = 90$. *The simple interest is* \$90.

6 *Any nonzero number raised to the zero power equals* 1. *So* $(-2)^0 = 1$.

7 $7.1 \times 10^2 = 710$; $7.1 \times 10^0 = 7.1$; $7.1 \times 10^{-2} = 0.071$. *Greatest to least: C.*

8 $2.8 + 7.2 = 10$, *so* $10 \times 10^5 = 1 \times 10^6$.

9 $m = \frac{5-(-1)}{2-(-4)} = \frac{6}{6} = 1$.

10 $m = \frac{10-4}{2-0} = 3$. The y-intercept is 4. So $y = 3x + 4$.

11 For any input x, there is exactly one output $y = -4x + 9$. Having a negative slope or negative outputs does not disqualify a function.

12 $8x - 6 = 42 \Rightarrow 8x = 48 \Rightarrow x = 6$.

13 $P(0) = 5$. For Function Q, $b = 9$. Since $9 > 5$, Function Q has the greater initial value.

14 $1^3 = 1$, $2^3 = 8$, $3^3 = 27$, $4^3 = 64$. The differences $(7, 19, 37)$ are not constant, confirming nonlinear.

15 The slope $m = 50$ represents the amount added per week, so the weekly deposit is \$50.

16 Graph B shows a steep increase (walking quickly), a flat section (waiting), then a gentler decrease (walking slowly home). Graph A's increase and decrease have similar steepness, so it doesn't match "quickly" then "slowly."

17 A rotation turns (spins) a figure around a fixed point called the center of rotation.

18 Two circles with the same radius are always congruent — one can be translated to overlap the other.

19 $180°$ rotation: $(a, b) \to (-a, -b)$. Reflect over x-axis: $(-a, -b) \to (-a, b)$.

20 When $k = 1$, every point stays in the same place. The image is congruent (identical in size and shape) to the original.

21 Alternate interior angles formed by parallel lines and a transversal are equal: $72°$.

22 $h^2 = 15^2 - 9^2 = 225 - 81 = 144$, so $h = 12$ ft.

23 The right triangle has legs 8 and 6. Distance $= \sqrt{8^2+6^2} = \sqrt{64+36} = \sqrt{100} = 10$.

24 A cylinder is 3 times the volume of a cone with the same base and height: $100 \times 3 = 300$ cm^3.

25 Lateral SA $= 2\pi rh = 2(3.14)(7)(10) = 439.6$ cm^2.

26 A curved pattern is a nonlinear association. The data has a trend, but it is not a straight line.

27 $y = 1.5(4) + 2 = 6 + 2 = 8$.

28 $x = 50$ is far outside the data range (1 to 10), so this is extrapolation.

29 Students who do NOT play an instrument are much more likely to play a sport (75% vs. 37.5%). This difference in conditional percentages signals an association.

30 With replacement: $\frac{6}{10} \times \frac{6}{10} = \frac{36}{100} = \frac{9}{25}$.

Practice Test 5 — Answer Key

1 Rational | 2 B | 3 A | 4 B | 5 C | 6 1 | 7 C | 8 C | 9 B

10 $y = 3x$ | 11 Yes | 12 D | 13 7 | 14 B | 15 A

16 The speed is constant for 4 seconds. | 17 A | 18 C | 19 C | 20 A

21 25 | 22 B | 23 A | 24 B | 25 B | 26 C | 27 Slope $= -1$; y-intercept $= 12$ | 28 A

29 Totals: Grade A = 30, Not A = 30, HW Yes = 40, HW No = 20, Grand Total = 60. Relative frequencies: $\frac{24}{60} = 0.4$

0 B

💡 Time to Learn! 💡

Review the explanations below, **especially for the questions you missed.**

Understanding why each answer is correct builds stronger problem-solving skills.

Tip: Circle any questions you got wrong, then read their explanation carefully.

Practice Test 5 — Detailed Explanations

1 $\sqrt{144} = 12$ exactly, because $12^2 = 144$. Since 12 is an integer, $\sqrt{144}$ is rational.

2 The repeating block has 2 digits, so multiply by $10^2 = 100$. Then $100x = 63.6363\ldots$ and $x = 0.6363\ldots$. Subtracting: $100x - x = 63$.

3 $4^2 = 16$ and $5^2 = 25$. Since $16 < 22 < 25$, we have $4 < \sqrt{22} < 5$.

4 Area $= \sqrt{20} \times \sqrt{8} = \sqrt{160}$. $\sqrt{160} \approx 12.65$ cm^2. (Note: $\sqrt{160} = 4\sqrt{10} \approx 12.65$.) Choices A and C confuse addition with multiplication.

5 Discount $= 120 \times 0.25 = 30$. Sale price $= 120 - 30 = \$90$.

6 $\frac{12^8}{12^8} = 12^{8-8} = 12^0 = 1$.

7 Move the decimal 3 places right: $6.1 \times 10^3 = 6{,}100$.

8 $\frac{2\times10^{-6}}{10^{-6}} = 2$. The bacterium is 2 micrometers long.

9 $m = \frac{7-3}{\frac{3}{2}-\frac{1}{2}} = \frac{4}{1} = 4.$

10 *Through the origin:* $b = 0$. *Slope* $= \frac{21}{7} = 3$. *Equation:* $y = 3x$.

11 *Every input* $(3, 4, 5, 6)$ *maps to exactly one output* (8). *Repeated outputs are allowed, so this is a function.*

12 $f(0) = -2(0) + 10 = 0 + 10 = 10.$

13 *The initial value is the output when* $x = 0$, *which is* 7.

14 *For a function to be linear,* x *must have an exponent of* 1. *Since* x *is squared in* $y = 3x^2$, *the function is nonlinear regardless of the coefficient.*

15 *Slope:* $\frac{22-10}{5-2} = \frac{12}{3} = 4$. *Using* $(2, 10)$: $10 = 4(2) + b \Rightarrow b = 2$. *So* $y = 4x + 2$.

16 *A flat section means the output (speed) stays the same — the object travels at a steady speed for those* 4 *seconds.*

17 *Add the translation to each coordinate:* $(1 + 3, 2 + 1) = (4, 3)$.

18 *Both reflection and translation are rigid transformations. Any sequence of rigid transformations preserves congruence.*

19 *The rule* $(x, y) \to (-y, x)$ *is a* $90°$ *counterclockwise rotation around the origin.*

20 *Congruent figures are a special case of similar figures with* $k = 1$. *All congruent figures are similar, but not all similar figures are congruent.*

21 $(2x + 5) + 3x + (x + 25) = 180$. *Combine:* $6x + 30 = 180$, $6x = 150$, $x = 25$.

22 $7^2 + 24^2 = 49 + 576 = 625 = 25^2$. *Since* $a^2 + b^2 = c^2$*, it is a right triangle.*

23 $d = \sqrt{(10-2)^2 + (9-3)^2} = \sqrt{64+36} = \sqrt{100} = 10$.

24 $V = \pi r^2 h = 3.14 \times 16 \times 10 = 502.4\ cm^3$.

25 *The lateral surface area (the curved side only) of a cylinder is* $2\pi rh$.

26 *The independent (explanatory) variable goes on the* x*-axis; the dependent (response) variable goes on the* y*-axis.*

27 $m = \frac{4-10}{8-2} = \frac{-6}{6} = -1$. $10 = -1(2) + b$, *so* $b = 12$.

28 $35 = 4x + 7$, $4x = 28$, $x = 7$.

29 *Add rows and columns for totals. Divide each cell by* 60 *for relative frequency.*

30 $P(A\ and\ B) = 0.6 \times 0.4 = 0.24$.

Practice Test 6 — Answer Key

1 Rational | 2 C | 3 B | 4 B | 5 B | 6 D | 7 B | 8 A | 9 A

10 A | 11 B | 12 C | 13 B | 14 B | 15 A | 16 B | 17 $(-4, 2)$ | 18 False

19 B | 20 30 cm | 21 B | 22 Yes | 23 C | 24 C | 25 B | 26 B | 27 A

28 B | 29 B | 30 C

Time to Learn!

*Review the explanations below, **especially for the questions you missed**.*

Understanding why each answer is correct builds stronger problem-solving skills.

***Tip:** Circle any questions you got wrong, then read their explanation carefully.*

Practice Test 6 — Detailed Explanations

1 $\frac{5}{6}$ *is a fraction of two integers with a nonzero denominator, so it is rational. Its decimal is* $0.8\overline{3}$*, which repeats.*

2 *The repeating block* 123 *has* 3 *digits, so you multiply by* $10^3 = 1000$.

3 $10^2 = 100$ *and* $11^2 = 121$. *Since* $100 < 110 < 121$*, we have* $10 < \sqrt{110} < 11$.

4 $\sqrt{a} + \sqrt{b} \neq \sqrt{a+b}$. *You must compute each root separately:* $\sqrt{4} = 2$ *and* $\sqrt{9} = 3$*, so the sum is* 5. $\sqrt{13} \approx 3.6 \neq 5$.

5 $I = 2000 \times 0.06 \times 2 = 240$. *Total* $= 2000 + 240 = \$2{,}240$.

6 *A negative exponent means reciprocal:* $4^{-3} = \frac{1}{4^3} = \frac{1}{64}$.

7 $\frac{8\times10^9}{3.3\times10^8} = \frac{8}{3.3} \times 10^1 \approx 2.4 \times 10 = 24$.

8 $870{,}000 = 8.7 \times 10^5$ *and* $4.3 \times 10^4 = 0.43 \times 10^5$. *Sum:* $(8.7 + 0.43) \times 10^5 = 9.13 \times 10^5$.

9 $m = \frac{1-7}{5-2} = \frac{-6}{3} = -2$.

10 $m = \frac{11-2}{4-1} = 3$. Using $(1, 2)$: $2 = 3(1) + b$, so $b = -1$. Equation: $y = 3x - 1$.

11 $y = 7$ is a horizontal line. Every x-value gives the same output 7, so it is a function. $x = 5$ is a vertical line (not a function). The other choices fail the vertical line test.

12 On the graph, go to $x = 2$ and read the y-value. The point is at $(2, 5)$, so $f(2) = 5$.

13 $|-4| = 4 > |-1| = 1$, so Function B decreases faster.

14 Check the differences in y: $8 - 5 = 3$, $11 - 8 = 3$, $14 - 11 = 3$. Constant change means linear. The others have non-constant changes.

15 Slope: $\frac{-1-(-4)}{1-0} = 3$. y-intercept: -4 (from $x = 0$). So $y = 3x - 4$.

16 Standing still means distance doesn't change: horizontal line. Then walking at a steady pace means distance increases at a constant rate: line sloping upward.

17 Subtract 6 from x and add 3 to y: $(2 - 6, -1 + 3) = (-4, 2)$.

18 A 3×4 rectangle and a 2×6 rectangle both have area 12, but they are not congruent because their side lengths differ.

19 A dilation by factor k from the origin gives $(kx, ky) = (2 \times 3, 2 \times 4) = (6, 8)$.

20 Perimeter scales by the same factor: $45 \times \frac{2}{3} = 30$ cm.

21 Corresponding angles are in the same position (e.g., both upper-left) at each intersection of the transversal with the parallel lines.

22 $11^2 + 60^2 = 121 + 3600 = 3721 = 61^2$. Since $a^2 + b^2 = c^2$, it is a right triangle.

23 Both points are on the x-axis. Distance $= |7 - (-5)| = 12$.

24 $V = \pi r^2 h = \pi(49)(3) = 147\pi\ m^3$.

25 $SA = 2(40) + 2(30) + 2(12) = 80 + 60 + 24 = 164\ ft^2$.

26 As x increases, y increases. The dots go up-right, indicating a positive association.

27 $m = \frac{8-3}{5-1} = \frac{5}{4}$.

28 Extrapolation means using the model to predict values outside the data range, which can be unreliable.

29 The 20 is the count for a specific combination of two variables (7th grade AND soccer), making it a joint frequency.

30 An impossible event has probability 0 because it can never occur.

Practice Test 7 — Answer Key

1 $\sqrt{2}$ (or $\sqrt{3}$) | 2 C | 3 9 and 10 | 4 B | 5 \$42 | 6 A | 7 $\approx 3.1 \times 10^{12}$

8 A | 9 B | 10 C | 11 1 | 12 B | 13 B | 14 C | 15 $y = 6x - 3$ | 16 C

17 C | 18 C | 19 B | 20 $\frac{3}{2}$ | 21 B | 22 B | 23 5 | 24 B | 25 6 cm | 26 C

27 $y = -1.5x + 9.5$ (approximately) | 28 B | 29 B | 30 C

Time to Learn!

Review the explanations below, ***especially for the questions you missed.***

Understanding why each answer is correct builds stronger problem-solving skills.

Tip: *Circle any questions you got wrong, then read their explanation carefully.*

Practice Test 7 — Detailed Explanations

1 $\sqrt{2} \approx 1.414$ *is between* 1 *and* 2 *and is irrational because* 2 *is not a perfect square.* $\sqrt{3} \approx 1.732$ *also works.*

2 $\frac{36}{100} = 0.36$ *(terminates). The correct conversion uses* $99x = 36$*, giving* $\frac{36}{99} = \frac{4}{11}$. *The student treated* $0.\overline{36}$ *as if it were* 0.36.

3 $9^2 = 81$ *and* $10^2 = 100$. *Since* $81 < 83 < 100$*, we have* $9 < \sqrt{83} < 10$.

4 $2\sqrt{5} = \sqrt{4} \cdot \sqrt{5} = \sqrt{4 \times 5} = \sqrt{20}$. *Alternatively,* $2\sqrt{5} \approx 4.47$ *while* $\sqrt{10} \approx 3.16$. *They are not equal.*

5 *Markup* $= 24 \times 0.75 = 18$. *Selling price* $= 24 + 18 = \$42$.

6 *Using the product rule:* $3^4 \cdot 3^2 = 3^{4+2} = 3^6$.

7 $498{,}000{,}000 \approx 5 \times 10^8$ *and* $6{,}200 \approx 6.2 \times 10^3$. *Product* $\approx 5 \times 6.2 \times 10^{11} = 31 \times 10^{11} = 3.1 \times 10^{12}$.

8 $\frac{6.4}{3.2} = 2$ *and* $10^{9-6} = 10^3$. *Answer:* 2×10^3.

9 $m = \frac{12-3}{4-1} = \frac{9}{3} = 3$.

10 *Steepness is* $|m|$: $|1| = 1$, $|3| = 3$, $|-5| = 5$, $|\frac{1}{2}| = 0.5$. *The steepest has* $|m| = 5$.

11 *By the vertical line test, if a vertical line crosses a graph more than once, it is not a function. So a function's graph is crossed at most once.*

12 $f(0) = -1$ *and* $f(2) = 7$. *So* $f(0) + f(2) = -1 + 7 = 6$.

13 *Same slope means the lines go in the same direction; different* y*-intercepts mean they never cross. Parallel lines have equal slopes and different intercepts.*

14 $6 - 3 = 3$, $11 - 6 = 5$, $18 - 11 = 7$. *The* y*-differences are not the same, so the rate of change is not constant — nonlinear.*

15 *Substitute* $m = 6$ *and* $b = -3$ *into* $y = mx + b$: $y = 6x - 3$.

16 *A horizontal line means the output does not change. The height remains constant during that time period.*

17 *Moving every point the same distance in the same direction is a translation (slide).*

18 *In* $\triangle ABC$: $\angle C = 180 - 55 - 80 = 45°$. *Since* $\angle C$ *corresponds to* $\angle F$ *in congruent triangles,* $\angle F = 45°$.

19 *Multiply each coordinate by* $\frac{1}{2}$: $(\frac{1}{2} \times 8,\ \frac{1}{2} \times (-4)) = (4, -2)$.

20 $\frac{9}{6} = \frac{3}{2}$, $\frac{12}{8} = \frac{3}{2}$, $\frac{15}{10} = \frac{3}{2}$. *The scale factor is* $\frac{3}{2}$.

21 $180 - 54 - 54 = 72°$.

22 $6^2 + 8^2 = 36 + 64 = 100$, *but* $11^2 = 121$. *Since* $100 \neq 121$, *it is not a right triangle.*

23 $d = \sqrt{(7-3)^2 + (4-1)^2} = \sqrt{16+9} = \sqrt{25} = 5.$

24 $V = \frac{1}{3}\pi(4^2)(9) = \frac{1}{3}\pi(144) = 48\pi\ cm^3.$

25 $2(80) + 2(10h) + 2(8h) = 376.\ 160 + 36h = 376,\ 36h = 216,\ h = 6\ cm.$

26 A strong positive trend corresponds to an r-value close to 1. $r = 0.9$ indicates a strong positive association.

27 Using $(1, 8)$ and $(5, 2)$: slope $= \frac{2-8}{5-1} = \frac{-6}{4} = -1.5$. $8 = -1.5(1) + b$, $b = 9.5$. Line: $y = -1.5x + 9.5$.

28 In $y = mx + b$, the slope m represents the rate of change: how much y changes for each 1-unit increase in x.

29 Of 42 pizza lovers, 24 are boys. $\frac{24}{42} \approx 0.571$.

30 There are 13 hearts in 52 cards: $P = \frac{13}{52} = \frac{1}{4}$.

Practice Test 8 — Answer Key

1 C | 2 $\frac{81}{99} = \frac{9}{11}$ | 3 ≈ 6.7 | 4 D | 5 3 years | 6 B | 7 C | 8 A | 9 B

10 $x = 2.5$ | 11 B | 12 C | 13 6 hours | 14 Nonlinear | 15 $y = 3x$

16 The person saved less per month in months 4–6. | 17 $(0, -4)$ | 18 C | 19 $(-2, 5)$ | 20 A

21 $83°$ | 22 B | 23 D | 24 $288\pi\ in^3$ | 25 11 in | 26 C | 27 Approximately 2 to 2.25

28 C | 29 Boys: $\frac{20}{30} \approx 66.7\%$. Girls: $\frac{12}{20} = 60\%$. The percentages are close, so association is weak or none.

30 B

Time to Learn!

Review the explanations below, ***especially for the questions you missed.***

Understanding why each answer is correct builds stronger problem-solving skills.

Tip: *Circle any questions you got wrong, then read their explanation carefully.*

Practice Test 8 — Detailed Explanations

1 Every integer n can be written as $\frac{n}{1}$, making it rational. Not every square root is irrational (e.g., $\sqrt{4} = 2$).

2 Following the same pattern: $99x = 81$, so $x = \frac{81}{99}$. Simplify by dividing both by 9: $\frac{81}{99} = \frac{9}{11}$.

3 $6.7^2 = 44.89$ and $6.8^2 = 46.24$. Since 45 is very close to 44.89, $\sqrt{45} \approx 6.7$.

4 $\sqrt{3} \times \sqrt{12} = \sqrt{36} = 6$ and $\sqrt{6} \times \sqrt{6} = (\sqrt{6})^2 = 6$. But $\sqrt{3} + \sqrt{12} \approx 1.73 + 3.46 = 5.19 \neq 6$.

5 $I = Prt$, so $t = \frac{I}{Pr} = \frac{60}{400 \times 0.05} = \frac{60}{20} = 3$ years.

6 The star is at $\frac{1}{8}$. Since $\frac{1}{8} = \frac{1}{2^3} = 2^{-3}$, the answer is 2^{-3}.

7 $\frac{8 \times 10^{-5}}{4 \times 10^{-6}} = \frac{8}{4} \times 10^{-5-(-6)} = 2 \times 10^1 = 20$.

8 $(9.5 \times 10^{-18})(2 \times 10^6) = 19 \times 10^{-12} = 1.9 \times 10^{-11}$ g.

9 $\frac{50-68}{6} = \frac{-18}{6} = -3$°F per hour. The negative sign shows the temperature is decreasing.

10 Set $y = 0$: $0 = 4x - 10$, so $4x = 10$, giving $x = 2.5$.

11 Graph A is a circle and fails the vertical line test. Graph B is a parabola opening upward — every vertical line crosses it at most once, so it is a function.

12 $f(-3) = -(-3) + 8 = 3 + 8 = 11$.

13 $8x + 20 = 3x + 50 \Rightarrow 5x = 30 \Rightarrow x = 6$.

14 Check the y-differences: $2 - 1 = 1$, $4 - 2 = 2$, $5 - 4 = 1$, $7 - 5 = 2$. The differences are not all the same $(1, 2, 1, 2)$, so the rate of change is not constant. The function is nonlinear.

15 Slope: $\frac{12-3}{4-1} = \frac{9}{3} = 3$. Using $(1, 3)$: $3 = 3(1) + b \Rightarrow b = 0$. Equation: $y = 3x$.

16 A less steep upward section means a smaller rate of change — less money was added per month compared to the first three months.

17 A $180°$ rotation uses $(x, y) \rightarrow (-x, -y)$. So $(0, 4) \rightarrow (0, -4)$.

18 Applying the $90°$ CCW rule $(x, y) \rightarrow (-y, x)$: $R(0, 0) \rightarrow (0, 0)$, $S(4, 0) \rightarrow (0, 4)$, $T(0, 3) \rightarrow (-3, 0)$. This matches $R'S'T'$.

19 Multiply each coordinate by $\frac{1}{2}$: $(-4 \times \frac{1}{2}, 10 \times \frac{1}{2}) = (-2, 5)$.

20 Dilation by $k = 2$ from the origin multiplies each coordinate by 2: $A(1, 1) \rightarrow (2, 2)$, $B(2, 1) \rightarrow (4, 2)$, $C(2, 2) \rightarrow (4, 4)$, $D(1, 2) \rightarrow (2, 4)$.

21 Co-interior angles are supplementary: $180 - 97 = 83°$.

22 The Pythagorean Theorem states that in a right triangle, $a^2 + b^2 = c^2$ where c is the hypotenuse.

23 $d = \sqrt{(4-1)^2 + (7-3)^2} = \sqrt{9+16} = \sqrt{25} = 5.$

24 $V = \frac{4}{3}\pi(216) = \frac{864\pi}{3} = 288\pi$ in^3.

25 $2\pi(4)h = 88\pi$, so $8\pi h = 88\pi$ and $h = 11$ in.

26 When data points group closely together, it is called a cluster.

27 Using endpoints: $m \approx \frac{11-2}{5-1} = \frac{9}{4} = 2.25$. A slope around 2 is a reasonable estimate.

28 For each 1-unit increase in x, y increases by the slope (6). For 3 units: $6 \times 3 = 18$.

29 Compare the conditional relative frequencies. Since 66.7% and 60% are similar, the preference doesn't change much by gender.

30 $P(\text{not event}) = 1 - P(\text{event}) = 1 - \frac{3}{4} = \frac{1}{4}$.

Practice Test 9 — Answer Key

1 C · 2 B · 3 A · 4 C · 5 B · 6 D · 7 C · 8 3×10^7 · 9 A

10 $y = x - 2$ · 11 B · 12 C · 13 B · 14 B · 15 A · 16 B · 17 B · 18 C

19 A · 20 B · 21 B · 22 B · 23 C · 24 C · 25 B · 26 A · 27 A · 28 B

29 C

0 *Independent: flipping a coin twice (first flip doesn't affect second). Dependent: drawing cards without replacement (first*

> ### Time to Learn!
>
> *Review the explanations below,* ***especially for the questions you missed.***
>
> *Understanding why each answer is correct builds stronger problem-solving skills.*
>
> ***Tip:*** *Circle any questions you got wrong, then read their explanation carefully.*

Practice Test 9 — Detailed Explanations

1 $1.41421356\ldots$ *is the decimal expansion of* $\sqrt{2}$*, which is non-repeating and non-terminating, indicating an irrational number.*

2 *Let* $x = 0.444\ldots$. *Then* $10x = 4.444\ldots$. *Subtract:* $9x = 4$*, so* $x = \frac{4}{9}$. *Note that* $\frac{4}{10} = 0.4$ *(terminates), which is different from* $0.\overline{4}$.

3 $6^2 = 36$ *and* $7^2 = 49$. *Since* $36 < 45 < 49$*, we have* $6 < \sqrt{45} < 7$. *The student is correct.*

4 $\sqrt{2} \approx 1.414$ *and* $\sqrt{3} \approx 1.732$*, so their sum is approximately* 3.146. $\sqrt{10} \approx 3.162$. *Since* $3.162 > 3.146$*,* $\sqrt{10}$ *is slightly larger.*

5 *Year 1:* $500 \times 0.04 = 20$*, total* $= \$520$. *Year 2:* $520 \times 0.04 = 20.80$*, total* $= \$540.80$.

6 $7^5 \cdot 7^{-5} = 7^{5+(-5)} = 7^0 = 1$.

7 *A negative exponent of* -27 *means the decimal moves* 27 *places left, making the number very small. There are* 26 *zeros after the decimal point before the digits* 167.

8 $\frac{7.2\times10^{12}}{2.4\times10^{5}} = \frac{7.2}{2.4} \times 10^{12-5} = 3 \times 10^{7}$.

9 $m = \frac{7-(-2)}{3-0} = \frac{9}{3} = 3$.

10 $m = \frac{1-(-2)}{3-0} = \frac{3}{3} = 1$. *The y-intercept is -2. Equation:* $y = x - 2$.

11 *Each student has exactly one student ID, so each input (name) maps to exactly one output (ID). That makes it a function.*

12 $f(4) = 3(4) + 1 = 12 + 1 = 13$.

13 *Hiker A's rate of change is 3 mph. Hiker B's slope is 4, meaning 4 mph. Since $4 > 3$, Hiker B walks faster.*

14 *A linear function has a constant rate of change (slope). Nonlinear functions have a rate of change that varies.*

15 *Slope* $= \frac{10-2}{3-(-1)} = \frac{8}{4} = 2$.

16 *A steeper upward slope means faster increase. The morning section is steeper than the afternoon section, so the temperature rose faster in the morning.*

17 *The $90°$ CCW rule is $(x, y) \to (-y, x)$. So $(2, 0) \to (0, 2)$.*

18 *Two rectangles can have the same perimeter but different dimensions (e.g., 2×6 and 3×5 both have perimeter 16), so they are not necessarily congruent.*

19 $90°$ *CCW:* $(x, y) \to (-y, x)$. *So* $(0, -5) \to (5, 0)$.

20 *When a figure is dilated by factor k, its area is multiplied by k^2. So the new area is $20 \times 3^2 = 20 \times 9 = 180$ cm^2.*

21 *Alternate interior angles are equal:* $4x = 80$, *so* $x = 20$.

22 $c^2 = 6^2 + 8^2 = 36 + 64 = 100$, *so* $c = 10$.

23 $d = \sqrt{1^2 + 1^2} = \sqrt{2} \approx 1.41$.

24 *The volume of a cone is* $V = \frac{1}{3}\pi r^2 h$, *which is* $\frac{1}{3}$ *of a cylinder with the same base and height.*

25 $SA = 2(9 \times 4) + 2(9 \times 2) + 2(4 \times 2) = 72 + 36 + 16 = 124$ *cm*^{2}.

26 *Clusters are groups of points that bunch together, while outliers are isolated far from the trend. Both can appear in the same scatter plot.*

27 *The line goes from about* $(0.5, 6.5)$ *to* $(7.5, 1.5)$. *Slope* $\approx \frac{1.5-6.5}{7.5-0.5} = \frac{-5}{7} \approx -0.71$.

28 $y = 5(8) + 20 = 40 + 20 = 60$ *dollars.*

29 *Two-way tables deal with categorical data (counts and proportions). Mean scores involve numerical data, not categorical.*

30 *Independent events don't affect each other's probabilities. Dependent events do — the first outcome changes the conditions for the second.*

Practice Test 10 — Answer Key

1 B | 2 D | 3 B | 4 D | 5 B | 6 81 | 7 5.9×10^{-6}

8 *50 people per sq km* | 9 B | 10 C | 11 $0, 3, 6$ | 12 B

13 F: rate of change $= 2$; G: rate of change $= 3$. Function G is greater.

14 B

15 B

16 Decreasing

17 C

18 $(3, -4)$

19 A

20 C

21 B

22 B

23 15

24 B

25 B

26 Nonlinear — height increases in childhood, then levels off in adulthood.

27 C

28 C

29 A

30 A

Time to Learn!

Review the explanations below, **especially for the questions you missed.**

Understanding why each answer is correct builds stronger problem-solving skills.

Tip: Circle any questions you got wrong, then read their explanation carefully.

Practice Test 10 — Detailed Explanations

1 5 is not a perfect square, so $\sqrt{5}$ is irrational and belongs in the Irrational region.

2 In Step 5, the student divided 2 by 10 instead of 9. From $9x = 2$, the correct answer is $x = \frac{2}{9}$.

3 $7^2 = 49$ and $8^2 = 64$. Since $49 < 50 < 64$, we have $7 < \sqrt{50} < 8$.

4 $\pi \approx 3.1416$, $\sqrt{10} \approx 3.1623$, and 3.2 is exact. From least to greatest: $3.1416 < 3.1623 < 3.2$.

5 Commission $= 250{,}000 \times 0.03 = 7{,}500$. The agent earns $\$7{,}500$.

6 $A = 3^3 = 27$ and $B = 3^1 = 3$. So $A \times B = 27 \times 3 = 81$.

7 Move the decimal 6 places right: $0.0000059 = 5.9 \times 10^{-6}$.

8 $\frac{3\times10^8}{6\times10^6} = 0.5 \times 10^2 = 50$ people per square kilometer.

9 A line that goes down from left to right has a negative slope.

10 Through the origin, $b = 0$. Slope $= \frac{12}{3} = 4$. Equation: $y = 4x$.

11 The inputs are the values in the left column of the mapping diagram: 0, 3, and 6.

12 Set $6x - 4 = 0$. Add 4: $6x = 4$. Divide: $x = \frac{4}{6} = \frac{2}{3}$.

13 From the graph, F rises from $(0, 4)$ to $(2, 8)$: slope $= \frac{8-4}{2-0} = 2$. Function G has slope 3. Since $3 > 2$, G has the greater rate of change.

14 A linear function is written as $y = mx + b$, where m is the slope and b is the y-intercept.

15 Starting amount $b = 120$. Spending \$15/day means losing money, so $m = -15$. The function is $y = -15x + 120$.

16 A graph that goes downward means the output (gas) is getting smaller over time.

17 Reflections are rigid transformations that preserve side lengths, angles, area, and perimeter.

18 Reflecting over the x-axis keeps x the same and negates y: $C(3, 4) \to C'(3, -4)$.

19 First translate: $(1 + 3, 4 + (-2)) = (4, 2)$. Then reflect over the y-axis: $(4, 2) \to (-4, 2)$.

20 *When $0 < k < 1$, the dilation is a reduction. $k = 0.25 = \frac{1}{4}$, so each length becomes $\frac{1}{4}$ of the original.*

21 *The three interior angles of any triangle always add to $180°$.*

22 *$a^2 = 17^2 - 8^2 = 289 - 64 = 225$, so $a = 15$. (Note: $\sqrt{225} = 15$, so B and D are the same value, but B is simplified.)*

23 *$d = \sqrt{9^2 + 12^2} = \sqrt{81 + 144} = \sqrt{225} = 15$.*

24 *$V = \frac{4}{3}(3.14)(125) = \frac{4}{3}(392.5) \approx 523.3\ in^3$.*

25 *$6s^2 = 150$, so $s^2 = 25$ and $s = 5$ cm.*

26 *Height rises quickly in youth, slows, and plateaus in adulthood. This is a nonlinear pattern.*

27 *A good line of best fit has roughly half the points above and half below.*

28 *$y = 2(4) + 1 = 9$.*

29 *$\frac{24}{80} = 0.30$.*

30 *With replacement: $P(\text{red}) = \frac{5}{8}$, $P(\text{blue}) = \frac{3}{8}$. $P = \frac{5}{8} \times \frac{3}{8} = \frac{15}{64}$.*

Well done checking your answers!

Keep practicing to strengthen your skills.

uthor's Final Note

hope you enjoyed this book as much as I enjoyed writing it. Whether you are a student working through the material, parent supporting your child's learning, or a teacher guiding your class, I have tried to make this book as clear and ngaging as possible. I hope I have succeeded. If you have any suggestions for improvement, please let me know. I ould love to hear from you.

ne accuracy of calculations is very important to me. We have done our best, but I also expect that I have made some inor errors. Constant improvement is the name of the game. If you find any errors, please let me know. I will fix em in the next edition.

***or students:** Your learning journey does not end here. I have written a series of books to help you learn math. Make ıre you browse through them. I especially recommend workbooks and practice tests to help you prepare for your kams.*

***or parents:** Thank you for investing in your child's education. I encourage you to explore the companion resources vailable online to help support your child outside the classroom.*

***or teachers:** Thank you for the invaluable work you do every day. I hope this book serves as a useful resource in your assroom. Feel free to reach out if you have suggestions or would like to discuss how best to use this book with your udents.*

also enjoy reading your reviews. If you have a moment, please leave a review on where you found this book. It will elp others find this book. If you have any questions or comments, please feel free to contact me at Nazari@ViewMath.com.

nd one last thing: Remember to use online resources for additional help. I recommend using the resources on `ttps://ViewMath.com` *You can find video lessons, practice problems, and more. You can also use the online ompanion for this book to track your progress and access additional resources.*

ishing all students the best in their studies, parents every success in supporting their children, and teachers ontinued inspiration in their classrooms!

r. A. Nazari

www.ingramcontent.com/pod-product-compliance
Lightning Source LLC
LaVergne TN
LVHW082348200926
848481LV00012B/3496